Funding Renewables: Powering Change

Michael

Table of Contents

1. Introduction

The energy policy shaping space is influenced by the demands of sustainability goals, associated with the complexity of the socio-economic and biophysical systems. Rising fossil fuel prices, societal impacts of energy use and observed climate change make the exploration of renewable energy more important than ever (Kowalski *et al.*, 2008). This paradigm shift, combined with the considerable amount of infrastructure and investment required for renewable energy, has led to an increasing presence of private and institutional investors including insurers, exposed to various risks.

According to the US Energy Information Administration (EIA), world energy consumption will grow by 48% between the period 2012-2040 (US EIA, 2017: online). This trend is attributable to strong economic growth emerging from Non- Organisation for Economic Cooperation and Development (OECD) countries (US EIA, 2017: online), including South Africa. Working as an alternative to fossil fuels, renewable energy technologies will play an integral role in meeting the world's growing energy demand adding to the global energy mix (IRENA, 2016:14). Furthermore, the share of renewables in the global energy mix can be cost-effectively doubled by 2030 using existing technologies. Along with energy conservation practices, an overall improvement in the global energy profile can be achieved which would significantly reduce greenhouse gas emissions and put the world on track to limit global mean temperature rise to below 2° Celsius (IRENA, 2016:14). This level of renewable energy deployment means scaling up current investment in renewables to 500 billion US dollars (USD) per year by 2020. Global investment must then reach USD 900 billion each year up to 2030 (IRENA, 2016:14). Almost two-thirds of this investment would be in the power sector, but renewables for heat and transport also need to grow significantly. Developing markets with fast growing energy demand will require the largest increase in investment (IRENA, 2016:14)

The South African energy sector underwent a shock, signified by a shortage in bulk electricity which was declared a national emergency in 2008, and in 2014. Eskom underwent other disturbances in energy supply when in 2015 it implemented 99 days of load shedding causing a decrease in mining and manufacturing output, as well as in subsequent years when sporadic load shedding was a norm. Leading to the outages was a shortfall in generation capacity and reliability problems with supply on the national grid and rapidly increasing costs. South Africa was to be tested by the compared stability levels of the 1990s, which were characterised by a 40% electricity reserve margin. During this period, South Africa's GDP grew steadily at 3.5% per annum and, as this trend persisted, no new capacity was added to Eskom, the South African electricity public utility responsible for generation, transmission and distribution. Albeit sourced from carbon intensive fossil fuel, the gradual decline in the reserve margin, had reached dangerous levels to the extent that the slightest interruption (such as routine maintenance) would compromise the integrity of the supply system, resulting in unplanned outages (Trollip *et al.*, 2014:8).

These risks as well as others are outlined in terms of the South African context particularly. Overall, the value in the book is its standing as a contribution to the understanding and subsequent informed and effective navigation of risks involved in the renewable energy sector in South Africa. Relatedly, the book addresses the gap in research on renewable energy and its implementation in South Africa.

1.1. South African context

In the description of South Africa, the mention of "1994" refers to the era when the first democratic elections, open to all citizens of voting age, took place. The year 1994 was a watershed time in South Africa's political history and symbolised the design and development of what was to be called "the new South Africa" or the "rainbow nation". The ANC won an indisputable 62.6% of the votes cast (Maharaj, 2008:8), and the election of the same year established a new social contract founded on the principles of social reconstruction, development, democratisation, transformation, reconciliation and nation-building (Kotze, 2000).

Before this period, South Africa was characterised by a protracted period of conflict between its civilians where divisions were sown between the races, with individuals classified as White enjoying political participation and competition, and non-whites (Blacks, Coloureds and Indians) enjoying only limited to no political rights. This system was termed *apartheid*, meaning 'separation', officially adopted by the National Party (NP) in 1948 (Seo, 2008:2). The South African government was considered a 'racial oligarchy', a regime where political power and economic privilege is consigned to a small fragment of society (in this case, the white minority). This consequently led to the isolation of the state from global economic participation in the form of sanctions (Seo, 2008:2). The government's separatist polices related to all forms of life, including an inferior education system aimed at preparing non-whites for subordinate positions in the workplace (Henrard, 2002:20). Although the apartheid system was legalised in 1948, it must be considered that it was built on a legacy of 200 years of white colonial rule, where racial exclusion and economic exploitation were promulgated for successive generations by Dutch and British colonists. Organised resistance to colonialism and apartheid was led by the African National Congress (ANC), which was formed in 1912 and was banned by the apartheid government in 1960. Apartheid was declared a crime against humanity by the United Nations in 1973 (Dugard, 1973: online), and a large-scale international campaign was developed to isolate the South African government and support the ANC and other liberation movements.

It took more than four years from the date of Nelson Mandela's (a political activist and leader of the ANC) release from the Robben Island prison in 1990 to negotiate the democratic transition. These negotiations involved the NP, at that time the incumbent government; the ANC, representing the majority of Blacks, Coloureds and Indians; and the Inkatha Freedom Party (IFP), representing the rural blacks of the KwaZulu region. Agreement was reached in creating an interim outline, called the Interim Constitution, which was to spell out the broad framework of the new democracy (Inman, 2013:2). It is against this political landscape that the country's energy sector has been shaped, and continues to evolve.

1.2. Research problem

This book explores the relationship between political risk, foreign direct investment, and the exploitation of renewable energy. In an analysis of the South African context, the book shows how South Africa's history of segregation as well as its current political profile have overall, affected foreign direct investment in the renewable energy sector. To make this point, the book draws on the Albert Venter (2005) model for political risk which takes into account, among other factors to be discussed and applied, the extent of political corruption, political policies such as Black Economic Empowerment, social upheavals, and the nature of macro-political and economic circumstances.

The book asserts that while the REIPPPP has produced positive results in the advancement of a renewable energy path in the South African energy mix, there are however obstacles that lead to a curtailment in the continuous and robust implementation of the Programme which prevents it from its full realisation. Since a robust renewable energy sector has been shown to attract foreign direct investment (as has been the case in China)South Africa's restrain to it represents a loss of opportunity to promote and attract direct foreign investment (FDI) and thus foster economic growth.

As will be demonstrated, there are three primary reasons for South Africa's restrained full adoption of renewable energy: firstly, the coal-dependent nature of the South African energy sector and the central position of coal to the South African economy means that any shifts away from this model would yield unemployment and challenges to the economy. These, though, would be in the short-term and the risk of a move toward renewable energy would yield long-term benefits such as heightened employment and increased FDI. Secondly, the drive for nuclear energy in conversation with Russia, which as will be shown is primarily driven by corruption (as shown in Section 4.5.5) and agendas not in the interest of the economy at large, has led to the stalling of meaningful and long-term renewable energy implementation. While various policies and papers on renewable energy have been drafted, and while some areas of South Africa (the Northern Cape particularly) have seen advances in the sector, successful renewable energy implementation falls victim to corruption and ulterior motivations. Thirdly, South Africa's history of racial segregation and the historical imbalanced provision of electricity has posed natural challenges for equal distribution of and access to it. In other words, access to energy in its most traditional form is already challenging for the government, and so pioneering an entirely different system may be even more so. Related to this is the structural issue of the clash between engaging in the global economy on the level of FDI and investing in developing the local community, especially in respect to disadvantaged communities who lack capital and access to electricity.

Overall, South Africa's positioning as a candidate for FDI and RE is hampered by factors relating both to government practices and decisions as well as to products of apartheid, and South Africa's history of racial segregation and imbalance. Attracting FDI through the implementation of a robust RE sector is further hampered by acts of corruption, as well as the dominance of the coal industry and the undesirable but almost necessary Eskom monopoly which yield structural challenges that are difficult to navigate.

This book investigates these complex factors in South Africa's political risk profile and RE sector with the intention of showing that political risk, FDI, renewable energy, and economic growth are intertwined concepts. The book integrates these three interrelated concepts and realities in application to South Africa in an exploration of its potential to attract FDI in the RE sector and, relatedly, the state of its renewable energy sector.

1.3. Rationale

Worldwide, extensive research and development (R&D) has produced many RE technologies to maturity ready for widespread application. However, the perception of the associated risk of RES and evidence of politically motivated interference, has constrained the progress of its scalability (Noothout, 2016: online). Unlike the renewables markets in South Africa, Eskom, a state-owned enterprise responsible for electricity generation, transmission and distribution, is the sole power purchase agreement (PPA) buyer, holding a monopoly that increases susceptibility to political intervention (Robson, 2013: online). This risk according to de Jongh *et al.* (2014) is prominent in the energy sector, as energy is the basis of any stable economy. Further, political involvement by governments is used to manage volatility in the economy.

Notably, the energy system is complex and is exposed to a variety of risks, whether operational, social, commercial, or natural. In globalised economies such as in South Africa where there is an open exchange of trade, renewable energy sectors are exposed to the kind of complexities that may be best navigated by using instruments such as a thorough political risk analysis. By considering the above-mentioned categories of risk that highlight some uncertainties of long-term effects, it is possible not only to identify particular risk factors that may hinder the full scale of benefits of RES, but also to capture the mechanics and intricacies of a domestic environment – the legal framework, social codes, and economic conditions thus increasing overall societal acceptance towards achieving success. The book uses an industry-accepted political risk analysis framework as a tool to study the risks involved in the promotion of renewable energy investments in South Africa.

1.4. Research questions

The South African government, through the REIPPPP, has produced positive results in the advancement of a renewable energy path in the country's energy mix. There are however obstacles that lead to a curtailment in the continuous and robust implementation of the Programme which prevents it from its full realisation. A strong renewable energy sector has been shown to attract foreign investment (as has been the case in China), and South Africa's restrain to it represents a loss of opportunity to promote and attract direct foreign investment and thus foster economic growth.

In view of this, the book asks:

- *What are the inherent political risks pertaining to the progression of South Africa's renewable energy sector?*

Sub questions are:

- *What has been the role of institutions in the transition to renewable energy exploitation?*
- *How have legacy features in South Africa impacted policy shaping in the energy sector?*

The book is qualitative in nature. It uses the Venter (2005) model as a tool for determining the relationship between political risk and RES as well as the use of existing academic work on political risk and energy as a means to contextualise these key areas.

1.5. Limitation of this study

The book is a desktop study which refers to the most recent market exploitation of a RES industry in the country. Market exploitation refers to the development of new knowledge about the firm's existing markets, products, and abilities (Norazlina *et al.*, 2013:120). Pertaining to South Africa's recently developed RE sector through the 2011 REIPPPP programme, there is a limit of academic papers available. In a few instances the author has selectively made use of grey and non-peer reviewed literature.

This book was predominately written in the first half of 2017, and subsequently there have been developments in the South African renewable energy sector which are not reflected.

1.6. Organisation of the book

This book consists of 5 chapters. The first chapter is the introduction, and here the context for renewable energy is set, as explained in the rise in price both monetary and societal cost of fossil fuels, highlighting the value of the sustainability goal. Information is provided on the political climate in South Africa and how it has emerged to influence the energy policies in the country. This is followed by a rationale for the book, and the questions that the book seeks to answer. Chapter 2 presents a comprehensive analysis of risk as captured in the discipline of political risk analysis and applied in the acquisition of foreign direct investment in the renewable energy sector. The chapter interrogates the concept of uncertainty, the basis of all risk, discusses how a country's political risk forms, and the relevance of its' analysis. A contextual discussion of the development of South Africa's energy sector provides a landscape of the political risks currently observed, concluded by a discussion of the most notable policies in the creation of the country's energy sector. Chapter 3 presents the methodology of a political risk analysis, and how the chosen methodology will be applied for the renewable energy sector. Chapter 4 uses the chosen methodology, and under five broad topics untangles factors that are impacting on South Africa's energy sector. Chapter 5 draws from the findings, providing conclusions and recommendations.

2. Literature review

In chapter 1, the concept of political risk is introduced as a growing interest builds between the intersection of politics and the globalisation of investment practices. It does this by demonstrating the relevance of political risk given the South African unique context, and the country's evolving energy sector.

In this chapter, an account for political risk is made as a study which is multidisciplinary, which can be understood from a number of dimensions. Political risk has a wide epistemology that traces back to founding theories, and evolving ones, that explain the formation of risk, how it presents, and the various methods used in order to capture it. Included in this discussion is a historical narrative of South Africa's energy sector and the entrenched minerals energy complex, including notable policies thereof.

2.1. Conceptualising Risk

Risk arises from uncertainty. However, while uncertainty suggests threat, there has been a paradigm shift from the idea that risk is associated exclusively with negativity: the newer trend suggests that risk can also represent measures of opportunity. This implies that risk can provide a prospect for change for the better (Frei & Ruloff, 1988:2).

The term "risk" in itself is broad and can be abstract and difficult to operationalise. Vlek & Stallen (1981:27) point to the following definitions as frameworks that can be used to understand risk:

- The possibility of loss;
- The size of the possible loss;
- A function and mostly the product of probability and size of loss;
- The variance of the probability distribution of all possible consequences of a risky course of action;

- The semi variance of the distribution of all consequences, taken over negative consequences only, and with respect to some adopted reference value; and
- A weighted linear combination of the variance of and the expected value of the distribution of all possible consequences.

The abovementioned definitions of risk present the various ways in which one can understand and capture risk. In summary, the above formulations show that risk usually refers to some or other loss, even though there is also scope for gain.

Mazareanu (2007:42) indicates that it is essential to identify exactly what is at risk, including the importance of the related vulnerabilities, with the aim to reduce the associated risks. Risk is influenced by the probability that an event will depend on a series of external factors. While external factors are comprised of the social, technological, economic, environmental and political spheres, internal factors focus on elements such as the historical data for the entity for which the risk assessment is made.

Ultimately, risk assessment is a preventive action. Corrective action, on the other hand, is taken through a disaster recovery plan. As risk is on the border of philosophy and mathematics, there are three domains within which risk can be classified – the real, the possible, and the impossible – all of which point to the essence of probability (Mazareanu, 2007:43).

2.1.1. Contextualising political risk

The scope of the definition of political risk is broad. Before considering the concept of political risk, this section unpacks the concept of politics itself.

Generally, politics refers to how a society is organised in terms of the laws of the land and its prevailing ideals. McKellar (2010:5) describes politics as a phenomenon that occurs at three main levels: global, regional or local. Social organisation can be influenced at these levels through official power. Dealing with political risk entails an interest in issues that exist in the political environment and an assessment of their impact for investors through the analysis of that information. It involves dealing with all relevant political events, under conditions of certainty or uncertainty, or with perfect or imperfect information (Kobrin, 1979:19). Ultimately, investors are concerned with the actions within the political sphere that may disrupt investment activity and profitability. Such actions include the decisions of government leaders with respect to tax laws, wage levels and environmental regulations.

The term 'political risk' is used frequently in business, investment and political economy literature. In application, 'risk' is usually equated to unwanted consequences. These consequences are usually of a political nature, where governments intervene in or affect business operations or investment opportunities. Examples range from a failure of a government to manage environmental instabilities emanating from violence, expropriation and taxation injustices, to public sector competition impeding on the profitability or objectives of the investment (Kobrin, 1979:68).

Similarly, to the concept of risk itself, *political* risk occupies a number of dimensions. Also like the broad concept of risk and the study of it, the study of political risk is multidisciplinary in nature, given the diversity of its application. In other words, political risk is heterogeneous in its epistemology and finds relevance between disciplines rather than within a singular discipline. For instance, in studies of international business, political risk will usually capture the concern of the management of exogenous factors that can influence market conditions (Jarvis, 2008:2). These exogenous factors refer to political decisions that can have an adverse effect on investor interests. Such factors include: government policies (fiscal, monetary, labour or industrial) that may asphyxiate the operations of a given investment thus creating transfer risk or risk to capital payments; operational risk, or business continuity; and ownership risk (Jarvis, 2008:3).

For the political scientist, political risk is concerned with the exercise of power, and consequently, harm that could be incurred by individuals, populations, non-states, and the international system. Political risk also refers to the factors that may hinder the smooth operation of political institutions, the exercise of legitimate rule and the functioning of the domestic and international society. Political risk is often regarded as political instability that is manifested through civil disobedience, low government legitimacy, poor public administration, and general state failure at the domestic level. At the international level, political risk is represented by diplomatic hostility, mercantilism and trade wars (Jarvis, 2008:6).

The study of political risk considers the maturity and transparency of the state to administrate and balance the competing demands of different constituencies. In this regard, aspects such as the independence of institutions such as the judiciary, financial and economic bodies of national accounts, as well as the electoral system, become key features to assess. Political risk in this context pays attention to economic actors, to the extent that corrupt administration is measured in terms of its cost to business, investment and economic growth (Jarvis, 2008:6). A country's political risk profile, in other words, would determine the extent of Foreign Direct Investment in, for example, the Renewable Energy Sector, which could, as noted, effect economic growth.

Political risk on the micro level, affects companies that carry high sunk costs such as energy and natural resources companies where they are threatened mostly by a subtle type of expropriation whereby tax rates are increased and invariably decrease the level of revenue, and the long-term profitability of the business (Sottilotta, 2013:10).

On the macro level, political risk concerns itself with the activities of governments and their impact on the interests of investment stakeholders. Change in any political parameter relating to policies that have a consequence for stakeholders (economic and non-economic) can thus be defined as political risk. The purpose of understanding and managing political risk relates not only to attracting FDI but also to the following tangible desirable outcomes: building institutional capacity; improving public administration; promoting transparency; increasing the efficiency and delivery outcomes of public administration to provide the necessary framework for economic growth; appraising various bodies on policy outcomes and inaction; helping business organisations to manage their exposures in foreign political environments; and preparing for, or mitigating the consequences of government action and public policy (Jarvis, 2008:6).

Over and above this, Calverly (1985:163) lists the following additional key areas for measuring political risk:

- Possible major currency devaluation;
- Major recession;
- Major shift in economic policy;

- Government pressure on multinational companies;
- Political unrest affecting economic performance; and
- Domestic banking failures

The areas listed above represent a broad approach that has been adopted by investors from developed countries. Overall, Calverly's (1985:163) list shows that political action and the state of politics in any country is intertwined with both global and local economics. Additional risk areas include: loss of copyright protection; discriminatory taxation; war damage; breach of contract due to political reasons; inconvertibility of currency or profit; and limits to remittances (Howell & Chaddick, 1994:73). Political risk emanating from the abovementioned sources has a substantial effect on industry and the growth prospects for a country. Efforts in these areas to promote stability and predictability result in the perception of a country as an attractive investment destination.

As far as acquiring information pertaining to country risk is concerned, Heinrichs & Stanoeva (2013: online) suggest a number of indices for calculating a country's risk that can be obtained from public sources. These are: the Corruption Perception Index (which ranks 170 countries according to the perception of corruption in the public sector); the Doing Business rankings (which monitor ease of doing business); the Global Competitiveness Index (which analyses 100 economic indicators to establish the competitiveness of the economy of a country and factors that determine the level of productivity of a country); the Gini Coefficient (an income inequality metric measuring the distribution of family income in a country); the UN Human Development Index (or HDI, which measures the achievements of countries in terms of health); and the World Bank Political Risk Indicator.

Understanding the intermestic nature of political risk is critical given the interconnection of states. Globalisation promotes a functional interdependence among states, where the increasing standardisation of international norms competes with the sovereignty of states and their societies. The impact of this may be, as Kobrin (1979:23) sets out to explain, that what may appear to be economic nationalism to an investor (through certain restrictions of FDI) may also be regarded as an attempt by the host country to implement a policy of indigenous industrialisation.

Kobrin (1979:11) also reminds us that all investors interact with all aspects of their environment (such as economic, political, sociocultural, legal and physical), with the distinction between these aspects existing on the experiential level. This suggests that the political system of an environment has much to do with the economics of that environment. This can be seen in the way that politics largely determines the framework of economic activity through economic policy, in that a society can see the result in a change from a market to a socialist economy or the reverse. Similarly, an economic event can be motivated by the need to maintain the power of the administration so as not to alienate important figures or interest groups (Kobrin, 1979:13).

The links between politics and the economy are apparent in the South African context where the end of apartheid compelled the government to establish a new economic policy that would promote a more equal society through strengthening democracy. The Reconstruction and Development Programme (RDP), a neo-liberal programme that ran from 1996 to 2000, put emphasis on its extensive welfare component. The RDP was abandoned and replaced by a series of other programmes as it proved unable to solve the prevailing problems of unemployment and the unequal distribution of wealth (Sherman, 2010:online).

2.2. Uncertainty

When it comes to political risk evaluation, it is generally accepted that the treatment of uncertainties in risk assessments, which is used in strategic decision-making, can be approached from different perspectives. Accordingly, the tools that are used provide useful support for decision-making because their outcomes inform the decision-maker's choice (Aven & Zio, 2010:64). Commonly, the foundation that has been used in risk evaluation is 'probabilistic risk analysis' (PRA), which dates back to the early 1970s. PRA is a systematic approach to risk mitigation that synthesises knowledge and uncertainties by addressing three fundamental questions: 1) which sequence of undesirable events transform the hazard into an actual damage? 2) What is the probability of each of these sequences? 3) What are the consequences of each of these sequences? (Aven & Zio, 2010:64).

This framework promotes knowledge of the problem and related uncertainties, which will be systematically manipulated by rigorous and repeatable probability-based methods to provide a clear representation of the risk outcomes (Aven & Zio, 2010:66). As the political risk levels of a firm operate as a component of its environment, it becomes important to establish the extent to which political events that occur in the environment impact the firm. The answer to this question lies in the understanding or evaluation of the relationship that exists between the firm and the environment, and "the relationship between events and outcomes" (Kobrin, 1978:14).

Concerning the relationship between events and outcomes, it is presumed that where there is a single outcome with a given event, or when a single outcome dominates all others, *certainty* exists (Kobrin, 1978:16). Under conditions of certainty, the multinational investor concerns him or herself only with determining the effect of political events on the magnitude of cash flows.

However, the idea of *objective uncertainty* would suggest that certainty does not exist, which means that the possibility of a single outcome does not exist, but also that one possesses perfect knowledge of all possible outcomes relating to the event and the probability of each occurring (Kobrin, 1978:16). Under conditions of objective uncertainty, the contribution of political events to business risk is a function of only the events themselves. From the above assumption, the multinational decision-maker would concern him or herself with the impact of politics on the expected value of cash flows and their distribution. The extent of risk is marked by the distribution of the joint probability of a political event taking place and influencing cash flows (Kobrin, 1978:22).

Finally, *bounded subjective uncertainty* suggests that all possible outcomes and their probabilities are unknown, and only opinions to their relative likelihood are available (Kobrin, 1978:22). Uncertainty in this formulation is subjective, meaning that there is neither knowledge of all possible outcomes, nor objective probabilities (Kobrin, 1978:22). The implication for the energy system here is that the contribution of political risk events is a function of both the event and the subjectivity factor (or the nature of perception). In this instance, stakeholders are concerned with the effect of political risk events on expected or desired outcomes and risk. The difference, however, is that risk is heightened as a result of inevitable subjective factors and the distortion that these may cause.

While the varying degrees or formulations of uncertainty explain the extent to which risk is present or is perceived, Kobrin (1978:17) reminds us that the *effect* of political events upon returns is provisional. This means that uncertainty is a function of both the environment and the firm. Risk itself is a property of the firm and the environment. Risk is expressed as the variation of a firm specific variable, from its expected value and can be *caused* by environmental events.

2.3. What is a Political Risk Profile and Why Does It Matter?

A political risk profile is a depiction of a single country's state of political risk. It is determined through employing certain criteria to measure the existing levels of risk. The importance of the evaluation and development of a political risk profile emerges given the following:

Firstly, classifying an investment by the country origin, and its associated characteristics (political, economic, social etc) acts as a useful step in identifying a group of investments that are likely to have similar characteristics and common sources of uncertainties, for instance, a type of climate, or population growth and labour trends. Secondly, the importance of developing a political risk profile is based on the existence of nation-states. All investments within a single state belong to the same government jurisdiction. Consequently, policies promoted by the government play a role in determining the profitability of the investments. This is to say that country risk analyses must consider the political policies and developments of the respective state (also known as sovereign risk).

These two considerations deal with exogenous risk, or risk that is beyond the control of the investor. Political risk analysis is seen to reduce the investor's subjective uncertainty about a large number of investments. The more favourable the outcome of this analysis, the more likely the investor will be to transfer capital that will equate to the rate of return across borders (Herring, 1983:81) and ultimately contribute to economic growth.

Political risk and the evaluation thereof are not a recent practice, and can be traced back to merchants, political leaders and military personnel. It has grown in prominence due to the exponential rise in foreign direct investment across the globe. This flow is seen as necessary to sustain economic development. A negative analysis may make the terms of negotiation unfavourable for the recipient country, over and above reducing initial interest in investment. As McKellar (2010:4) puts it, *"the one variable that effects international business operations that is exogenous, difficult to control and potentially hazardous is, in fact, political risk"*.

2.4. Formation of a country's political risk

According to Simon (1984:127), the formation of political risk involves the interplay between the conditions that prevail in the host country, home country environment and global environment. Host-country and home-country conditions refer to: government policies; societal attitudes; local business community actions; legal community rulings and media reports; while international country conditions refer to the foreign policies (economic, military, diplomatic) of nation-states; regional organisations' policies; international activist groups policies and internal developments in nation-states. In contrast, the global environment refers to global organisation policies (such as those of the UN and IMF) and global developments (such as worldwide inflation, the oil crisis and external debt crises).

The formation of risk can be the result of direct external risks, such as protests in the investing (home) country against investments in certain states, home government restrictions on overseas operations, and attempts to monitor the operations of the organisation and impose codes of conduct. This interplay has the ability to strain and damage relations between the host and home countries (Simon, 1984:127).

Additional factors that may influence the formation of political risk profiles include the stage of economic development and the degree of openness in the socio-political systems of the home and host countries.

2.5. History of South Africa's energy sector

South Africa's energy sector is characterised by a heavily capital and energy-intensive development pathway, based almost entirely on coal. This pathway has been driven by resource extraction and the development of a connected set of interrelated economic activities termed the 'Minerals Energy Complex' (MEC), which was primarily based on mining, limited beneficiation, and linked industries and underpinned by the provision of some of the cheapest electricity in the world. Eskom has been the cornerstone of the MEC and, in turn, the Complex has become central to the economy (Greenpeace, 2012).

Coal thus historically played, and continues to play, a central role in the South African economy. The consideration of coal and energy is therefore integral to a political risk analysis of the country – because energy functions as a key part of South Africa's economy, the state of it would directly and indirectly affect the extent of FDI. Illustrating this is coal as a source of economic value, contributing R51 billion to South Africa's economy in 2013 (StatsSA, 2015: online). With a mining value add of 22.5%, the mining industry contributed 8% to the gross domestic product (GDP) in 2016. The size of the mining workforce in 2015 was estimated at 480 146 individuals with the platinum group metals (PGM) industry making up at the largest workforce at 41%, followed by gold at 21% and coal at 20%. At 77% of total primary energy supply, it is also the country's dominant energy source (StatsSA, 2015: online).

Since the 1970s, Eskom (established in 1923 and named Escom until the 1980s) had massively expanded electricity-generating capacity through the construction of several large-scale coal-fired power plants, growing installed capacity from 6500 MW in 1969 to over 25000 MW in 1990 (Burton & Winkler, 2014:1). However, efforts that went to this development were racially biased and imbalanced, as revealed in an apology at the Truth and Reconciliation Commission in 1997 (TRC, 1997: online). According to the TRC apology, the Electricity Act, which granted local authorities sole control over electricity, supplied only within their areas of jurisdiction, causing the separation of naturally integrated networks. Separating the administration of black urban areas from white cities often meant that the black areas were left without electricity services. Eskom initially took no positive action to broaden access to electricity and it did not challenge racial policies that prevented investment in and development of townships and outlying areas or homelands (TRC, 1997: online).

Currently, the electricity generation mix is made up by coal (91%) supplied by state owned monopoly Eskom, by means of 13 coal-fired power stations with an installed capacity of 37,745 MW. Eskom also has 1,800 MW of nuclear, 1,332 MW of hydro/ pumped storage, 2,426 MW Open Cycle Gas Turbine (OCGT) and 103 MW of wind installed, for a total installed capacity of 45,389 MW (Fisher & Downes, 2015). Coal exports provide a substantial source of foreign revenue, accounting for R50.5 billion in 2011 (South African Coal Roadmap, 2013:1). The dominant participants in coal production, accounting for 80% of coal production in the country, are known loosely as the major five: Anglo-American, Exxaro, Sasol, BHP Billiton and Xstrata (Eberhard, 2011:3).

The historical evolution of coal in South Africa can be traced to its industrial application and the birth of the mining sector in the 19th century. Mining activities were centered on diamond mining, and then gold upon the discovery of this metal in the 1890s in Johannesburg. Consequently, coal-generated electricity became the principal source of energy for mining and minerals processing, as well as the supporting infrastructure surrounding the mining economy (Marquard, 2006:72). Most of this coal mined in South Africa comes from the Central Basin, which was discovered in 1920 and includes the Witbank, Highveld, and Emerlo coalfields. However, coal from this area is projected to be exhausted by the next century (Du Plessis & Henry, 2006).

Three key features have shaped the South African coal industry. In the 1920s, coal mining was subject to grading, and the price of coal was subject to market forces, which were underpinned by the influential behaviours of the cartels in the sector (Marquard, 2006:75). The prominent cartels that shaped policy-making, monitoring uncompetitive practices, and regulation were the Transvaal Coal Owners' Association (TCOA), the Natal Associated Collieries (NAC), the Anthracite Producers' Association, and the Coke Producers Ltd Mining. An important feature of the cartels was their control over the price of coal (ANC Policy Institute, 2012:47). This measure promoted exports and made South African coal and energy price the cheapest across these countries (ANC Policy Institute, 2012:47). By the 1940s, coal was regularly used by private consumers (for direct use, such as heating); the railway system, which was responsible for transport to distribute coal to the consumer, and itself relied on steam coal as fuel; an emerging electricity sector; and for export (Marquard, 2006:75). Production was thus labour-intensive yet characterised by unsafe practices.

The second key feature of the coal industry was the state setting price controls on the domestic market and limiting exports during the period 1940-1970. The method used to calculate local prices was based on a rate of return, which excluded the cost of depreciation. This method was chosen in order to promote industrialisation and combat inflation, resulting in South Africa's energy prices remaining globally one of the lowest. However, the desired effect of being competitive on the global front was met by a fracture in the international coal market, culminating in South Africa's exclusion from the global economy owing to mounting anti-apartheid pressure, along with an apparent substitution of coal by oil. South Africa's coal sector was beset with a labour-intensive production structure, low mechanisation, low extraction levels, and little product differentiation and beneficiation. Furthermore, capital investment from the old mines was low and growth in the coal market was subject to long term contracts from Eskom, where coal prices were set below wholesale price (Marquard, 2006:76).

The third key feature, which characterised the period from 1970 to present, was entrenched in a policy called the Petrick Report. This document constituted real increases in the regulated domestic price of coal and introduced increased extraction and utilisation rates as a means of issuing export permits. The effect of such a program was a massive expansion of the electricity industry, as well as a massive penetration of the synthetic fuels industry (Marquard, 2006:76).

Still in the same period, expansion of the coal sector was then further fueled by Union action against the notoriously low labour costs. The effect of this revolt led to government introducing policies promoting mechanisation, domestic energy security, and increased exports. Both these events led to a large-scale increase in the demand for coal, and a consequent revitalisation of the export market, which in 1973 benefited from the oil crisis (Marquard, 2006:77). However, existing rail and port facilities presented a major logistical bottleneck. As part of resolving this, a condition with the Japanese government for a coking coal contract lead to the development of new port that could handle larger bulk carriers, resulting in the establishment of key infrastructures – the Richards

Bay Coal Terminal (RBCT) north of Durban and a dedicated rail link connecting the most important coalfields to the port – thus providing alleviation to the logistical challenges (Eberhard, 2011:7).

The prospects of the country's coal mining sector then began to change. Increased demand, higher regulated prices, and an energised export sector could be compared to a market capitalisation in the 1960s of R175 million to over R2 billion in the 1970s. The coal sector was thriving and experienced exponential output levels, which quadrupled by the end of the century, as well as investment in the form of increased production capacity with the emergence of new coalmines (Marquard, 2006:77). Furthermore, the high potential value of exports and the foreign exchange earned was seen as critical to the development of the South African coal mining industry, promoting more efficient extraction rates and possible cross subsidizing of cheaper, lower grade coal to the domestic market. The domestic market grew substantially as the national electricity utility embarked on a robust capacity expansion program in the 1970s and 80s. At the same time, Sasol, South Africa's only commercial synthetic fuels producer, also made incremental investments in coal to liquids (CTL) new plants (Eberhard, 2011:8).

The lead-up to the democratic elections of 1994 saw the introduction of reformist policies that exacted the official collapse of the apartheid regime and the entry of South Africa into the global economy. This came with trade liberalisation, which opened up the state's borders, raising South Africa's economic profile and attracting interest in the country from the global community. This sentiment was no different in the energy resource sector where trade liberalisations accelerated the internationalisation of South Africa's coal industry. Collieries, which had previously functioned as divisions within gold mining houses, were restructured into separate subsidiaries, each with their own local and overseas marketing operations (Eberhard, 2011:8). In 1999, Anglo Coal, for instance, commenced its global strategy, which included defending the profitability of its South African assets while expanding its footprint in other parts of the world (ANC Policy Institute, 2012:124). Furthermore, as much effort was made post-1994 and in much of the early 2000s to redistribute South Africa's wealth, the Mineral and Petroleum Resources Development Act (MPRDA) and other mining charter requirements were introduced to diversify corporate ownership structures of coal. These new structures were put in place to generate a new order of black capitalism and to promote different strategies to coal mining (ANC Policy Institute, 2012:135). The scope of mining thus changed as emerging minors came out of Black Economic Empowerment deals (a consequence of the post-apartheid legislations), with African Rainbow Minerals, Shanduka, and other small scale local mining companies playing a part (Burton & Winkler, 2014:2).

Meanwhile, a shortfall in generation capacity and reliability problems with supply on the national grid and rapidly increasing costs were emerging. South Africa's shortage in bulk electricity supply was declared a national emergency in 2008 and again in 2014. Eskom experienced other disturbances when unplanned outages in 2015 led to 99 days of load shedding causing a decrease in mining and manufacturing output, as well as in subsequent years when sporadic load shedding became a norm. Leading to the outages was a shortfall in generation capacity and reliability problems with supply on the national grid and rapidly increasing costs South Africa was to be tested by the compared stability levels of the 1990s, which held a 40% electricity reserve margin. During this period, South Africa's GDP grew steadily at 3.5% per annum and, as this trend persisted, no new capacity was added to the Eskom fleet. There was thus a gradual decline in the reserve margin, to the extent that the slightest interruption (such as routine maintenance) would compromise the integrity of the supply system, resulting in unplanned outages (Trollip *et al.*, 2014:8).

The unfolding of these events is captured in the 1998 White Paper on Energy Policy of the Republic of South Africa (hereafter referred to as White Paper 1998). In the same way that South Africa's energy policy was formally influenced by international relations (as seen in economic sanctions as mentioned in Section 1.1, including on exports, specifically coal), similar forces were shaping the country's choices as a post-apartheid and liberalised country participating in the global economy. The country was developing an energy policy of energy supply and use that resembled international trends, which implied ensuring energy security by means of diversification and flexibility of supply sources and energy carriers and relying on market-based pricing (see the distribution of electricity in Figure 2:1). This policy also suggested the inclusion of privatisation and competition, and the role of government in creating a policy that attracted investment while ensuring the achievement of national policy objectives (DME, 1998:9).

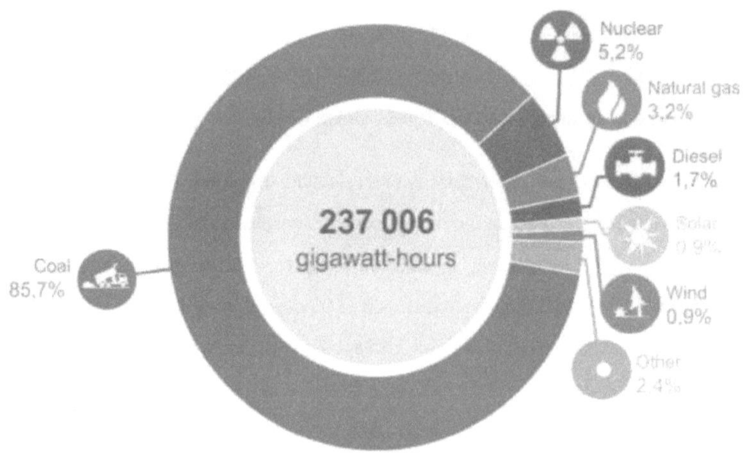

Figure 2: 1 Total electricity generated by source as at 2016. Source: StatsSA (2018).

By 2012, Eskom was almost entirely dependent on coal, and sustained by demand from mining, industrial, agricultural, commercial, and residential customers and a historic mandate that drives economic development through the investment of large-scale centralised electricity generation (Koen, 2012:5). The current situation of energy output is now expected to increase from approximately 44 000 MW to 80 000 MW between 2012 and 2030 to meet the projected growth and demand (KPMG, 2014:32). The concern of South Africa occupying a high position in the top CO_2 emitters globally, the government defined strategies to increase power generation capacity as part of a commitment to drive down the country's energy intensive, coal-based profile. Such strategies were outlined in the Integrated Resource Plan (IRP) in 2010 and an (intended) updated Plan in 2013. The Plan aims to change the country's electricity generation mix from high coal dominance (85%) to a moderate one (48%), with prominently high levels of renewable energy. Furthermore, as the country's electrification rate of 85% still leaves parts of the population without electricity, Eskom's Transmission Ten-Year Development Plan 2013-2022 states that the major focus of the new developments in the transmission network are to ensure that the new power stations developed by independent power producers (IPPs) are integrated, and that new loads are connected to the network (KPMG, 2014:32).

Foreign direct investment (FDI) in the energy sector, and specifically renewable energy stands as a desired outcome not only as a mechanism for climate change mitigation, but also for bridging the energy gap, and the generation of economic growth. National development policies such as the National Development Plan (NDP) and the Draft White Paper on Foreign Policy highlight the importance of FDI for meeting South Africa's development priorities in addressing the triple challenges of poverty, unemployment and inequality. While some determinants of FDI may be further from the control of government, such as natural resource endowments and a large domestic market, aspects such as a stable political and policy environment are within the reach of government and are attractive investment determinants.

2.6. Policy frameworks in the establishment of the renewable energy sector

2.6.1. The White Paper on Energy Policy 1998

The White Paper on the Energy Policy of the Republic of South Africa, published in December 1998, was drafted in conjunction with the National Electrification Programme. This was a programme implemented between 1994-1999 with the mission to fast track electricity roll out to black South Africans in underdeveloped urban and rural areas. Drafted in post-apartheid South Africa, it is symbolised by a democratically elected ANC dispensation. The White Paper (1998) energy objectives are the following: 1. to Increase access to affordable energy services with reference to poor households, small businesses, small farms, and community services. 2. To improve energy governance by strengthening the consultative process in the formulation and implementation of energy policies and having clear delineation of roles and responsibilities amongst institutions 3. To stimulate economic growth and create an investor friendly environment where there is competition in the energy markets, and relatedly, to implement a transparent regulatory framework in conjunction with strong governance structures. 4. To manage energy-related environmental and health impacts by balancing the exploitation of fossil fuels and placing measures that avert the negative health impacts resultant of certain fuels used by poor households, and 5. To secure energy supply through diversity (DME, 1998:8-9).

The White Paper (1998) also introduces the political mandate for the unbundling of electricity generation, transmission, and distribution. The focus of the Paper was on introducing competition into the energy sector, particularly through the involvement of the private sector, with emphasis on the importance of allowing energy consumers the right to choose their electricity provider (DME, 1998). To this end, independent power producers (IPPs) would be allocated up to 30 per cent of South Africa's generation. The involvement of IPPs at the time of publication was in respect of building and connecting new coal-fired power stations to the grid, with little intention for renewable energy generation (Morris & Martin, 2015:23)

2.6.2. Renewable Energy White Paper 2003

The Renewable Energy White Paper (REWP) (2003) was developed with the purpose of diversifying South Africa's energy generation, effectively disrupting how Eskom operated, and to introduce the private sector in contributing to and diversifying South Africa's energy generation. The REWP derived its mandate from the Constitution and committed government to set a broad target of 10,000GWh of renewable energy by 2013, which entailed 4% of projected demand for 2013 mostly from small scale hydroelectricity, biomass, wind solar (Madhlopa *et al.,* 2014:16) and to develop a strategic framework and mechanisms to achieve that goal. Due to delays in the actual implementation of the policy, the target of 10,000GWh was not met by 2013 (Morris & Martin, 2015:24)

The REWP lists the following areas of focus and development:

1. Financial allocation and the setting of targets, fiscal incentives, providing financial support, of national resources towards renewable energy technologies, as well as creating the climate to attract foreign direct investment.
2. Creating legal instruments and a regulatory framework for pricing and tariff structures aimed at integrating IPPs into the existing electricity system.

3. Technology development- and advancing research and development that promotes renewable energy technologies, as well as developing guidelines and codes of practice for the suitable use of renewable energy technologies.
4. Governance in the form of decisive management of the renewable energy program and policy coordination with government departments creating an enabling environment.
5. Awareness raising and education- promoting knowledge in order to increase adoption and use, and improve communication amongst energy role players on energy polices (DoE, 2015: 18-19).

2.6.3. Integrated Energy Plan 2003

The first Integrated Energy Plan (IEP) was published in 2003 by the Department of Minerals and Energy[1] (DoE, n.d.: online). The IEP is a multi-faceted long-term energy policy that provides guidelines for energy policy decision-making and the analysis of different pathways of the South African economy up to 2020. This long-term policy considers features such as energy reserves, energy consumption, future generation capacity, possible energy mixes and technologies to meet demand, and the greenhouse gas (GHG) emissions of those possible energy sources. The IEP takes a holistic view to the country energy mix and considers job creation, minimizing the impact on the environment, and the use of water as a scarce resource, and coordinate it with national development plans and strategies, international commitments, and long-term climate change mitigation strategies (Madhlopa *et al.,* 2014:18). Since the IEP acts as the energy roadmap for South Africa, the National Energy Act of 2008 requires that an IEP be formulated on an annual basis and report on all energy consumption and energy generation over a 20-year horizon. In reality, this has not been the case. The second IEP report was drafted in 2012 by the DoE, became open for public comment in 2013, and scheduled for submission to the Cabinet by the end of 2014. The finalisation of the IEP is significant, not only in charting the future trajectory of South Africa's energy mix, but in laying the technical foundation for the official revision of the Integrated Resource Plan (IRP) (DoE, n.d.: online).

---

[1] On 10 May 2009 then newly elected President, Mr Jacob Zuma, announced the creation of the Department of Energy to replace the Department of Minerals and Energy (DoE, n.d.:online).

2.6.4. Electricity Regulation Act of 2006

This Act was developed by the Department of Minerals and Energy (DME), the objectives of the Electricity Regulation Act (ERA) are to:

1. Achieve efficiency of the electricity supply infrastructure in South Africa
2. Safeguard electricity users within a sustainable electricity supply industry- using energy regulation
3. Facilitate investment in the electricity supply industry
4. Facilitate universal access to electricity
5. Promote energy diversification
6. Promote competitiveness and diversification for the electricity consumers and
7. Balance the interests of end user licensees and investors (Electricity Regulation Act 2006:8).

A key outcome of the Act was the establishment of the National Energy Regulator of South Africa (NERSA), which came to substitute the National Energy Regulator (NER) and assigned to upholding the above-mentioned objectives. The NERSA acts within the confines of the IRP and is the ultimate entity that determines electricity tariffs; grants licenses for the generation, distribution and transmission of electricity; and controls the import and export of electricity. The ERA outlines NERSA's relationship with Eskom where the Regulator is given separation of powers and autonomy in its function. The ERA has furthermore had major influence on the IPP programme and the subsequent creation of the Renewable Energy Feed-in Tariff (REFIT) and REIPPPP.

The ultimate creation of energy policy however lies with the DoE, which stipulates that NERSA is to be given licensing power for new generation activities, and that Eskom is to be given purchasing power of any new generation. In effect, these specific roles force all three bodies to align in agreement with the allocation of new generation capacity for IPPs to be connected to the national grid. This division of mandates has influenced Eskom to muster its power as sole purchaser to delay the process of bringing private sector generation online (Morris & Martin, 2015:25). In 2016, Eskom refused to sign in excess of 40 IPPs onto the grid citing an electricity surplus. These projects relied on a guaranteed off-take from Eskom, and would inject 58 billion ZAR and potentially create 15 000 local jobs (Corkin, 2018:5).

2.6.5. Long-Term Mitigation Scenarios paper 2007

The Long-Term Mitigation Scenarios paper (LTMS) was developed by a Scenario Building Team, consisting of strategic thinkers from government, academia, business and civil society. They were tasked with the creation of a document that would play a significant role in South Africa's climate change policymaking as well as in advancing the country's renewable energy programme (Scenario Building Team, 2007:1-22). The aim of the LTMS is to provide a sound scientific analysis on South Africa's GHG emissions mitigation scenarios upon which Cabinet could draw up a long-term climate policy. Such a policy would give South African negotiators clear and delegated positions for their negotiations under the United Nations Framework Convention for Climate Change (UNFCCC). The policy would additionally act as a guideline to commit to a range of realistic strategies for future climate action. Through the scenario mapping team, starting from the base year of 2003 and continuing to 2050, two possible scenarios are created: that of no mitigation efforts before 2050 as part of "growth without constraints" and that of full-scale mitigation as "required by science". These scenarios are placed against a full range of possible international climate change contexts (Scenario Building Team, 2007:1-22).

The more robust second scenario presents the opportunity for the exploring of four strategic options for reducing greenhouse gas emissions. Strategic option one: ("start now") as an option is symbolic of energy efficiency efforts and points to mitigation actions that have upfront costs with the prospect of future savings. The option leads to a relative reduction in emissions, with an average of about 230 Mt CO2-eq avoided each year. Strategic option two ("scale up") involves using economic and regulatory instruments to scale up the efforts in "start now" by adding more positive cost actions, leading to total emission reductions of around 13 800 Mt CO2-eq between 2003 and 2050. Overall mitigation costs are equivalent to 0.8% of GDP (Scenario Building Team, 2007:1-22) which fall well below the 1% benchmark suggested by the Stern Review, and positively linked to consistent growth and development in a low-carbon energy economy (World Bank, n.d.:13). Strategic option three ("use the market") builds on the first two options to promote the accelerated adoption of efficient technologies and changing social behaviour using incentives and taxes. The option reduces emissions by 17 500 Mt CO2-eq between 2003 and 2050. Strategic option four ("reaching for the goal") cannot be modelled as it is based on new or unreleased technologies that still require extensive research and development (R&D) operation in the market (Scenario Building Team, 2007:1-22).

The LTMS informed then President of the Republic of South Africa, President Jacob Zuma's (in office from 2009-2018) to pledge at the Copenhagen Conference of Parties (COP) in 2009 to reduce GHG emissions by 34 per cent from business as usual by 2020. This pledge was ambitious and surprising since South Africa is a non-Annex 1 country under the Kyoto Protocol, and by definition is not required to implement caps on GHG emissions. The LTMS also served as the basis for future energy mix planning, informing both the IRP in 2010 and the National Climate Change Response White Paper in 2011 (Morris & Martin, 2015:26).

2.6.6. New Generation Regulations 2009, and Electricity Regulation Act 2006

The Electricity Regulation on New Generation Capacity was published under the Electricity Regulations Act, 2006, and published on 5 August 2009. Its objectives involve:

1. The formation of a regulatory framework to uphold the relationship between the buyer and an independent power producer (IPP) into a power purchase agreement (PPA) thus making it free of unfair practice,
2. The facilitation of a cost reflective electricity tariff, and the full recovery thereof by the buyer of all costs incurred through its relationship with the PPA,
3. Setting the framework of approving the IPP bid programme, and, before it was replaced, the REFIT programme.

This framework would facilitate the procurement process and further set the rules applicable in the undertaking of both the IPP and REFIT programmes (DoE, 2009:6). During the development of REFIT, the DoE initiated a process of revising the Regulations on New Generation soliciting the involvement of foreign consultants. This led to the formation of the Electricity Regulation Bill (Second Amendment), which came into effect in May 2011. The Bill contained several changes to the original Regulations on New Generation Capacity that effectively transferred powers that used to reside within NERSA and Eskom's mandate to the DoE. NERSA's power to formulate the PPA and selection criteria was removed. The selection process of the REFIT IPPs was taken out of Eskom's control and the bill made it possible for the Minister of Energy to instruct Eskom to buy power from an IPP. In effect, the Bill placed the development of an IPP programme under the control of the DoE and laid the framework for regulating the REIPPPP, the implementation of which was not previously supported by any existing policy (Morris & Martin, 2015:26).

2.6.7. Integrated Resource Plan (IRP) 2010, and the IRP Updated Report 2013

The Integrated Resource Plan for Electricity 2010-2030 (IRP 2010) (DoE, 2011), adopted in March 2011, lays out the country's proposed generation fleet for the 2010-2030 period. As a "living plan" (Montmasson-Clair & Ryan, 2014:6), it is informed by the IEP and its resultant scenarios. An update of the IRP 2010 was published in November 2013 for public comments. The prescription was that it should be revised every two years in order to mirror changes in the country's energy demand and supply profile (Montmasson-Clair & Ryan, 2014:6). This, however, has not been achieved.

Notably, the energy sector has undergone changes in its electricity demand outlook since the endorsement of the Integrated Resource Plan (IRP) 2010-30 due to changes in the economic climate, including the recession following the financial crisis in 2008. A revised economic and electricity sector outlook has been developed to inform decisions in the 2013 Update. The demand in 2030 is now projected to be in the range of 345- 416 TWh as opposed to 454 TWh expected in the policy-adjusted IRP. This is a reduction from 67800 MW to 61200 MW and a resultant 6600 MW less capacity required of generating capacity. The Update considers the aspirational economic growth of 5,4% per annum until 2030 suggested by the National Development Plan in order to reduce unemployment and alleviate poverty in South Africa. It also considers a reduction in electricity intensity in the economy, where economic growth does not grow at the same rate as electricity intensity. It is also aligned with a shift in the structure of the economy away from energy intensive industries which is assumed to dramatically reduce the electricity intensity of the economy allowing the growth rate to have a less imposing impact on electricity demand to 2030 and beyond (DOE, 2013).

With the Update, future generation options, like nuclear and CSP, are to be weighed against electricity demand, cost, and progress made on alternative generation technologies. For example, the Update Report delays decision-making on nuclear power generation until 2015 based on the conditions that electricity demand exceeds 270 TWh, there is no shale gas development, and nuclear generation costs are under US$6,500/kW (Morris & Martin, 2015:26). However, the DoE has yet to publish the IRP Update Report as the new IRP iteration. According to industry experts, the DoE's delay is most likely due to the Department's hesitation to place limitations on nuclear energy development given the considerable amount of support for nuclear development in government (Morris & Martin, 2015:26).

2.6.8. National Climate Change Response White Paper 2011

The National Climate Change Response White Paper (NCCR) was published in 2011, prior to South Africa hosting of COP17 in Durban. It sets out the following two objectives: 1. To create capacity in the economic, environmental, and social context to manage climate change impacts, 2. To make a fair contribution to the global effort to stabilise greenhouse gas (GHG) concentrations in the atmosphere at a level that avoids dangerous anthropogenic interference with the climate system.

The Paper incorporates the principles set out in the Constitution, the Bill of Rights, the National Environmental Management Act, the Millennium Declaration and the United Nations Framework Convention on Climate Change. It provides a broad overview of the country's climate change response framework by outlining priorities for adaptation and mitigation strategies (DEA, n.d.:5). It further includes plans that advance a transition of jobs from a carbon-intensive economy to a green economy under the banner "million climate jobs", a relevant feature where in South Africa jobs are a key priority (WWF, 2011:5).

2.6.9. Independent System and Market Operator (ISMO) Bill

Eskom is a vertically integrated power utility, owning all levels of the supply chain which are generation, transmission, and distribution. These three levels represent the areas of competition, given an enabling regulatory framework. Unbundling and deregulation of the electricity market involves splitting up these utilities into separate generation, transmission, and distribution companies and introducing an Independent System Operator (ISMO) entity (Steyn, 2013:540-548). An ISMO essentially would function autonomously from Eskom, usually to operate the national grid, fulfilling a planning, expansion and procurement function as well as a system operation function with regards to new generation capacity. The planning function is related to the modelling of scenarios contained in the Integrated Resource Plan (IRP), which acts as input to the transmitter for the development of the expansion plan based on electricity demand as per the IRP (Steyn, 2013:540-548).

The function of the ISMO would be to promote new private sector participants, where the generation sector IPPs are ensured fair access to the national grid as well as an offtaker for the electricity generated by them. The ISMO removes stumbling blocks, such as cases where access to the national grid could be retarded by the generator owner in favour of its own new developments, as well as obstructions, such as access to the transmission system, delayed connection to the transmission system, and unequal bargaining power for the sale of electricity. This is especially true given that the generator-owner has economies of scale and lower sunk costs than the IPP by virtue of its existing infrastructure (Steyn, 2013:540-548). The ISMO Bill was announced in 2011 by then President Jacob Zuma as "critical for the power supply of South Africa". In 2015 it was withdrawn with no explanation (Fin24, 2015).

3. Methodology

Chapter 2 introduces political risk as an element that captures the interplay between business- and in this case FDI which is placed in the energy sector, and politics. As will be discussed in Section 3.3 amongst other areas, FDI in RES has been a significant contributor to the growth of the economy highlighting the importance of understanding the sources of risk. This chapter discusses political risk analysis as a technique that assesses the likelihood of loss due to a political situation in a state. It does so by discussing methods used to investigate political risk, including how these methods have evolved over time.

3.1. Evolving Methodologies in Political Risk Analysis

Over time, various influences have shaped the way in which political risk has been operationalised. Though the approaches have evolved with time, the recognition of the cross-pollination that exists between politics and investment has been consistent.

The earliest approach to political risk can be traced back to the 1950s and the Catalogue School (Jarvis & Griffiths, 2007:11). Here, political risk is viewed as a consequence of actions taken by host governments that adversely affect the operations and profitability of investors. As a result, focus was placed on a list of negative activities of governments (or societal inertia) that hamper the functioning of efficient markets (Jarvis & Griffiths, 2007:11). A few flaws were identified in this approach, namely the assumptions that markets are perfect, prone to equilibrium, and self-regulating. In this approach, the possibility of imperfect markets, poor transparency, monopolistic tendencies or collusion were absent from consideration. The method presented a disjointed picture of foreign business operating as a separate entity outside of its socio-political surrounds, and as being affected negatively by state participation in business. The role of the state is disassociated from its role as an enabler of commercial practice and is instead viewed as a driver of greater levels of political risk (Jarvis & Griffiths, 2007:11). This method was criticised for its lack of "operational traction [...] and conceptual and analytical looseness" (Hudson & Leftwich, 2014:31).

The advent of second-generation approaches in the 1970s, such as the System–Event School, meant adopting a more intuitive approach to political risk by correlating certain types of states to certain levels of political risk. It was understood that economic growth was not indicative of low levels of political risk, and that there is a mutuality between political systems and markets. Second-generation approaches emerged during a period characterised by political modernisation theories, which were examining the decolonisation process, the democratisation of societies and the emergence of fragile states following the Second World War (Jarvis & Griffiths, 2007:13). The central concern here was the cause of the political development of societies to political development, and, relatedly, the reasons for political cultures and political institutions to support economic systems and production networks. Characteristic of the System–Event School was the postulation that political risk and political instability had a negative correlation with modernisation. Such an assumption is standard for developed, predominantly Western and capitalist economies. This view was widely criticised, as it ostracised developing economies based on their ethnic origins. It also proved untrue, as investors are known to operate successfully in certain African countries despite civil turbulence (for example, GlaxoSmithKline Pharmaceuticals, MTN, SABMiller in various parts of Africa). The System–Event School therefore was unable to demonstrate causality between political events and an impact on investments (Jarvis & Griffiths, 2007:15–16).

The 1980s were demonstrative of a shift in thinking in political risk analysis as captured by third-generation studies, under which attention was paid to developing nations and the political risks investors were met with in these contexts. This entailed developing specific methods to understand political power and wealth distribution and the effect thereof on societies (Hudson & Leftwich, 2014:33). This was during a time when expropriation, nationalisation and protectionist policies were on the decline, and governments were actively positioning themselves to compete for FDI and attract prospective investors. Third-generation approaches endeavoured to focus on micro-analyses and stressed the importance of specific context rather than grand theoretical correlations. This meant developing explanatory schemata and methods to evaluate the risk environment in relation to the investment. This led to the introduction of a plethora of qualitative techniques characterised by deductive reasoning, none of which were privileged over others (Jarvis & Griffiths, 2007:18).

The on-going evolution of political risk modelling seeks to address the drawbacks of those of earlier generations when faced with the challenge of providing data sets that can be used by analysts to depict the relationships between political and economic institutions, domestic norms and external forces. The 1990s and current period have produced fourth-generation methods which are pioneering the use of analytical precision in identifying relevant institutions in evaluating the interplay between politics and investment in a process that can add to public policy delivery and public sector resource efficiency. Here, a diverse range of methods attempt to develop systematic methodologies for identifying trigger points linked to various political risks associated with food security and famine, ethnic and religious tensions, civil conflict, inter-state hostilities, energy crises and environmental sustainability (Jarvis & Griffiths, 2007:19-21).

Though fourth-generation methods are overcoming weaknesses identified in previous schools of thought, they have not yet recognised indigenous approaches to political risk and the value of a local understanding and contextual nuance over the generally accessed information sources found in most political risk assessment models. These sentiments are partially summarised by Ripka (2007:93) who calls for the need to build up a "genuine" theoretical framework when examining political risk. Political risk analysis assembles information that offers insight and allows the client to understand the environment where the potential investment is headed.

Notably, Ripka, (2007:56) makes a distinction between the concepts of "forecasting theory", "theory of interaction" and the supply of information on "political" aspects. In terms of political risk assessments, the forecasting outcome must have a foundation on analysis that describes a state's past and current conditions. This description must depict the key factors that make up the workings of the political conglomerate while accenting the events that may induce losses, determined by the theory of interaction. In this way, what emerges is a systematic view of factors driving political risk that can better inform risk management strategies, or indeed the decision to invest.

The importance of a political risk analysis therefore lies in its ability to preserve the importance of the relationships among the variables in the assessment model. In using descriptive analyses of political risk in South Africa, emerges information about future sources of loss within the country.

3.2. Methods of investigation

Political risk analysis is a technique that assesses the likelihood of loss due to a political situation in a state. It does not, however, consider or measure outcomes that could be caused by economic circumstances or business cycles or practices. Political risk assessments usually come in the form of a letter grade or a numerical score from a state as a whole, with a written justification of that particular grade or score (Howell & Chaddick, 1994:72). The kind of risk portrayed in the report can either represent an opportunity or a potentially damaging risk, which could entail a gain or gamble for the investor.

According to Howell (1998:9), there are limitations to establishing a precise prediction, even with the use of various analytical methods. While a chosen model is designed to reveal situations that may bring harm to investor interests, social systems are notably complex, and social phenomena must often be simplified and abbreviated in the model. There is also the possibility that some of the representative variables are not necessarily the best choices. For instance, it may happen that certain variables are not included, and this will have some adverse effect on investment decisions, or alternatively, important variables may have been unforeseen or exist outside the scope of the analysis.

Though the research done to compile political risk models is meant to aid in-depth, descriptive analysis that offers explanatory insights for often complex and intricate political, economic and social phenomena, this may get lost in quantitative techniques, and there is always the possibility of incomplete and sometimes inaccurate information. Therefore, variation in the data will introduce some margin of error in projections, and there is also error created through human intervention (Howell, 1998:9).

Essentially, methodologies employed for a political risk analysis can be quantitative (employing statistical or mathematical operations) or qualitative (standardised checklists, Delphi technique, and scenario analysis) in nature, and can also combine subjective and objective approaches. Different methods may be used to assess the exposure to political risk that an investor possesses, and the methods employed depend on the size of the investor's firm, that firm's degree of internationalisation, and the type of industry (Anchor, 2012: online).

Combinations of more than one procedure per investigation may be employed in order to assist the analysis and render a forecast. According to Ascher & Overholt (1983:73), methods include the following:

- Extrapolation – the projection of a historically based quantitative trend at a constant or regular rate. This rate may be a particular numerical rate of increase or a particular rate of acceleration.
- Regressions – the prediction of one event or forecast based on its relationship with one or more other trends or variables.
- Leading indicators – with this method, the direction of change in the trend under examination is presumed to depend on changes in the direction of other trends. This is often used in economic forecasting.
- Multi-source forecasts – an amalgamation of several sources that involves the processing of other forecasts. It is the result of combining the results of expert interactions.

The type or combination of methods employed will generally be chosen based on various factors such as the availability of information, the size of the budget, the type of industry examined, and the form of information required.

3.3. A theoretical framework for the energy sector

There is often confusion over the concept and terminology associated with political risk. Whereas political risk *assessment* is a probability measure of future risk that acts as a warning signal of the level of threat, political risk *analysis,* the object of this book, considers the origins or causes of the threat (Howell, 2009:5). The aim of a political risk analysis is to concentrate on the theoretical aspect of political risk, and to offer a descriptive account of the levels of risk. This is often neglected in commercial political risk models, where variables are selected because they are topical, giving rise to bias in the development and application of the models (Howell, 2009:5).

Essentially, political risk captures the functioning of business and politics as two different but interrelating domains. Given that the purpose of business is to create profit for itself and its stakeholders, and to conduct business within the confines of a political authority and social organisation, business is naturally affected by politics and the laws of the state (McKellar, 2010:6).

In addition to the domestic nature of political risk analysis, there is an international element that contributes to the "intermestic" nature of the understanding of political risk. This includes the elements that affect cross-border relations, most notably in the area of investments. Investors' exposure to risk, which is associated with the transactions that take place between the many different states, has also become salient (Herring, 1983:75). In the 21st century globalised world, the international environment has an impact on shaping political risk within the country. Factors such as immigration, engagement in international trade and investment relations and policy development are all relevant here. In South Africa this process is particularly notable given the process of liberalisation fostered by the end of apartheid.

The post-apartheid period has been largely characterised by the formation of policies and the re-examination and re-construction of an energy policy inspired by the global trends of diversification and the flexibility of energy supply. One such feature of international trend and best practice is an energy sector characterised by market-based pricing with emphasis on commercialisation and competition. Such an energy system requires government to create a domestic environment that attracts investment, while ensuring the achievement of national policy objectives (DME, 1998:7). In other words, the nature and functioning of the government sector. Logically, then, political risk shares an inextricable relationship with energy concerns. This is particularly pertinent to South Africa where, as mentioned, the state-owned provider Eskom stands as the dominant energy provider in the country. As such, the activities of the government have a direct impact on the state and nature of the local energy sector and the perceived risks associated with it.

Studies on the determinants of FDI flows to developing countries, such as those of Busse and Hefeker (2005: online) argue that there is a lack of research concerning descriptive analyses of political dynamics and other relevant policies in host countries and the impact of these on FDI. Such a prospect can be dangerous for a country's energy sector, which is often cited as the bedrock of a stable and growing economy. Wafo (1998:5)states that "a clear understanding of the link between FDI and the political environment may understate corporate risk strategies, forgo opportunities, or the prospects of international capital flows that would otherwise add towards the growth of world trade". Many of the international studies that do exist for example the Political Risk Services (PRS) Group, are from developed countries and use narrow definitions of political risk that focus on specific indicators, such as democratic rights, property rights and trade policy. Resultantly, these do not fully capture the politics behind risk analysis. In addition to their use of narrow definitions and their exclusion of the broader aspects that constitute policy-related variables that render these studies unable to capture the nuances of political risk, they are also irrelevant to the contexts of developing countries such as South Africa. By contrast, the use of indigenous models, (such as the Venter (2005) model used as the framework for this book) in conjunction with an in-depth appraisal of the political climate as it relates to the energy sector, is a more effective strategy for determining the variables that impact the formation of an evolving energy sector.

A limitation relating to the application of the Venter (2005) model, however, is its inability to provide singularly defined potential risk events relating to *forecasts* that may be used for the processes of investing (Howell, 2009:8). Moreover, the model does not assign weighting to its constituent variables, therefore treating each of them as equal. The most useful application of this model, then, is to the *current* state of the energy sector.

As such, this analysis will make use of Venter's (2005) model as a South African approach to political risk analysis in the present. In developing his approach, Venter (2005) draws on Howell and Chaddick's (1994) identification of the three most influential approaches to political risk assessment: namely, the Economist Intelligence Unit (EIU), the Political Risk Services model (PRS) and the Business Environment Risk Intelligence model (BERI). Venter then uses these to configure a model for political risk analyses and assemble a model that is responsive to the South African context.

The Venter (2005) model is based on 15 topics that encompass political, economic, and social dimensions. Venter (2005) establishes 15 factors relevant to political risk. These are: threatening neighbouring states and foreign policy environment; authoritarian measures to retain power; staleness of incumbency and leadership succession; legitimacy of government; military involvement in politics; social risk (including terrorism and religious fundamentalism); socio-economic conditions; racial, ethnic and religious cleavages; black economic empowerment; trade union activism; safety and security; labour policy; macro-political and economic circumstances; administrative (in)competence in government; and the security of private property. As is evident, these 15 topics cover a wide range of areas. The commonalities or patterns that can be traced through them, though, can be grouped under the following topics: socio-political upheavals; efforts towards socio-economic change; overriding\existing conditions; government efficacy; and stability in the private sector. Table 3.1 shows the distribution of risk.

Table 3:1 Political risk variables

Variable	Description
1	Threatening neighbouring states and foreign policy environment
2	Authoritarian measures to retain power
3	Black economic empowerment
4	Trade union activism
5	Labour policy
6	Macro-political and economic factors
7	Socio-economic conditions
8	Racial, ethnic and religious cleavages
9	Staleness of incumbency and leadership succession
10	Legitimacy of government
11	Administrative (in)competence in government
12	Safety and security
13	Military involvement in politics
14	Social risk (including terrorism and religious fundamentalism)
15	Security of private property

☐ Socio-political upheavals ■ Efforts towards socio-economic change ☐ Overriding/existing conditions
☐ Interrogation of the efficacy of government ■ Stability in the private sector

The Venter model was last consulted in 2005, and again in 2014, meaning that the information in it (outlined below)represents a current revision of the method that depicts the county's political risk profile as recently as 2017, as it applies to a developing renewable energy sector and the FDI traction thereof. The purpose of the model is to formulate a political risk profile of South Africa that considers the relationship between the risks that constitute the profile and the ambitions of the country's energy sector. Drawing on this model, this book establishes the contribution of political risk towards FDI growth in the country's renewable energy sector. Such reflection is important given that FDI has been the primary source of investment in the renewable energy sector and that foreign lending has been identified as one of the key areas that are reliant on political risk and its analysis (Ripka, 2007:19).

The premium that states place on FDI as a source of economic growth has made it vital for countries to position themselves in a way that will attract it. This can be achieved *"by controlling political and economic variables, focusing on a country's pull factors, making intelligent choices when coming to policy, and publicising the worthiness of the country"* (Lewis, 2000:101).

Illustrating the significance of RE and its related FDI contribution to economic growth is the fact that between the periods of 2010-2015, R192.6 billion was invested in renewable-energy projects in South Africa. Of this, 28% (R53.2 billion) came from foreign investment and finance. This figure almost matches the entire foreign direct investment in the South African economy in 2014 (R62 billion) (Breytenbach, 2015:2). The concern of RES, and an in-depth understanding of the state of RES (part of which involves understanding the risks of its implementation) contributes toward the ultimate promotion of FDI. As such, the importance of this book lies in its concern with RES, political risk, and relatedly FDI as an integral contributor to the growth of the economy.

4. Findings and discussion

In chapter 3, a discussion is made on political risk analysis and the FDI that has emerged in the South African energy sector. In this chapter, the Venter (2005) political risk model is considered. Reference is made to its 15 dimensions and their impact on and relevance to the country's developing renewable energy sector and FDI growth thereof. Through this exercise, links between political risk factors and the energy sector will demonstrate the interconnectivity and reliance of these two worlds in the attraction of FDI growth in renewable energy.

In so doing, the political risk analysis examination will draw upon literature from the pre-1994 era in order to provide a context and a deeper understanding of where the country comes from, as well as to demonstrate the ideological evolutions, policies, economy, and social conditions, that have emerged to create the current landscape of South Africa.

Venter (2005) establishes 15 factors relevant to political risk. They can be broadly grouped into 5 categories, namely: socio-political upheavals; efforts towards socio-economic change; overriding\existing conditions; government efficacy; and stability in the private sector.

4.1. Socio-political upheavals

4.1.1. Foreign policy environment

After the ANC assumed power in 1994, there was a need to reform policies from the apartheid system in order to reflect a new philosophy of political inclusion aimed at nation-building. This was a period in South African history when it was also important to craft a foreign policy that would be responsive to international relations and engage with a globalising environment. In March of 1994, the ANC-led government published a comprehensive foreign policy document called "Foreign Policy Perspectives in a Democratic South Africa" (Le Pere, n.d.: online). This document was based on the values of upholding human rights, promoting of democracy and adhering to international law (Le Pere, n.d.: online). Furthermore, because of South Africa's liberation history, its international relations are based on two pillars: Pan-Africanism and South–South solidarity (DIRCO, 2011:5–6). This means that South Africa recognises itself as an integral part of the African continent and SADC and realises that part of its success is linked to Africa's stability, unity and prosperity. South Africa's foreign policy further commits "to the centrality of multilateralism, consolidating relations with the North, and the strengthening of bilateral social, political and economic relations" (DIRCO, 2011:5–6).

4.1.2. South Africa and authoritarian measures to retain power & military involvement in politics

As explained by Booysen (2013), power refers to the relationship between "control" and "consent" that governs the "ruler" and the "ruled". Political power refers to leaders/parties/governments that operate within the public sphere, muster public resources, and make decisions and take directions that will affect the greater population. This sort of control is fuelled by the electoral engine, of which voter support is the key driver.

Even though the South African government is based on a power-sharing or multiparty system, the ANC is dominant in its power over the state. While the dominance of the ANC is attained through a legitimate, democratic electoral system, Brooks (2004:2) warns that a dominant party system is unfavourable to democracy, as it can lead to a scenario where there is an erosion of office, and the dominant party responds less and less to public opinion, given an unofficial guarantee of re-election. This scenario means that South Africa fairs relatively poorly against the Albert Venter (2005) model for political risk in terms of measures to retain power. While South Africa by no means function as an autocracy, the sheer dominance of the ruling party does heighten its political risk in the respect of Venter's (2005) concern of (loosely) "authoritarian measures to retain power" and "staleness of incumbency". In other words, in this criterion, South Africa would struggle to attract FDI on a significant scale.

Rivero (2014:7) further warned that if, under the administration of the Zuma presidency, the ANC government did not halt engagement with nepotism, abuses of political power and cadre deployment, it would continue to witness revolts such as the "brain drain", protest action, and increased voter apathy, which would all, as per Venter (2005), damage the country's capacity for attracting FDI. Notably, there has been something of a decrease in the ideological gap between the population groups in the country, and the race factor (which has been a notable point in favour of the ANC among the black population) is beginning to lose prominence, being replaced by class indicators (Rivero, 2014:7). Perhaps this is best demonstrated by the emergence of the Democratic Alliance (DA), a historically white-dominated political party, which is steadily gaining traction as the official opposition party, with support rising from 12.4 per cent in 2004, to 16.7 percent in 2009 (Waller, 2004) and now 22.2 per cent (Electoral Commission of South Africa, n.d.: online).

4.1.3. Implications of the above variables for the renewable energy sector

The Southern African Development Community (SADC) as a regional block constitutes 14 member states, namely; Angola, Botswana, Democratic Republic of the Congo, Lesotho, Malawi, Mauritius, Mozambique, Namibia, Swaziland, Tanzania, Zambia, Zimbabwe, South Africa and Seychelles. The region and its member states is made up of a mixture of uneven levels of socio-economic human of development stages. States like South Africa, Seychelles and Botswana for example possess developed economic infrastructures and markets, unlike those that are developing and grappling with complex socio-economic challenges (SADC, 2012:6). Unlike the European Union, most of which is connected in some respects institutionally and in other respects as a monetary union, SADC comprises of postcolonial states with governments of each state that sometimes contend with public opinion and citizens opposing a unified regional singular entity. As a result of these factors, the energy sector within SADC struggles to establish a unified and fiscally-sustainable region. For instance, in 2012 electricity deficits in the region were supposed to be addressed through the establishment of various plans and projects such as the Mozambique Backbone Project, the Central Transmission Corridor (CTC), the Zambia-Tanzania-Kenya Interconnector and the proposed Namibia-Angola Interconnector. However, to date, these projects have not materialised due in part to delays in project implementation (SADC, 2012:6).

In the 21st century the region has a low access to electricity: 24% compared to 36% for the East African Power Pool (EAPP) and 44% for the West African Power Pool (WAPP), with some of the

SADC countries having below 5% rural access to electricity. The energy challenge facing SADC occurs in the context of an unstable economic climate. Moreover, the 2008 financial market crisis had an impact on the attractiveness of the energy infrastructure for foreign and local investors. This also resulted in the continuous dominance of coal as the energy source of choice (SADC, 2012:6). Indeed, the coal industry is the backbone of power generation in the region with a significant share of the resource earmarked for export (Oosthuizen, 2014:9).

Domestically, the Constitution (1996) allots legal powers to different channels of government to advance an inclusive and just energy policy that sufficiently taps into the natural energy resource of the country catering to the needs of the country. The Constitution promotes the sustainable attainability and affordability of the production and distribution of energy and for it to reach citizens despite their geographic location, and thus improve their standard of living overall. Perhaps the most notable achievement in South Africa's energy sector has been the involvement of South Africa in the establishment of the Southern African Power Pool (SAPP) in 1993, which was signed by all SADC countries in 1995. The SAPP aims to provide electricity to all SADC countries in an environmentally sound manner (see the electricity mix in SAPP countries in Figure 2:2). It also focuses on drawing hydropower from the Inga Hydropower plant which has a potential of generating up to 100,000 MW using the natural flow of the Congo River (Department of Environmental Affairs and Tourism, 2005:33).

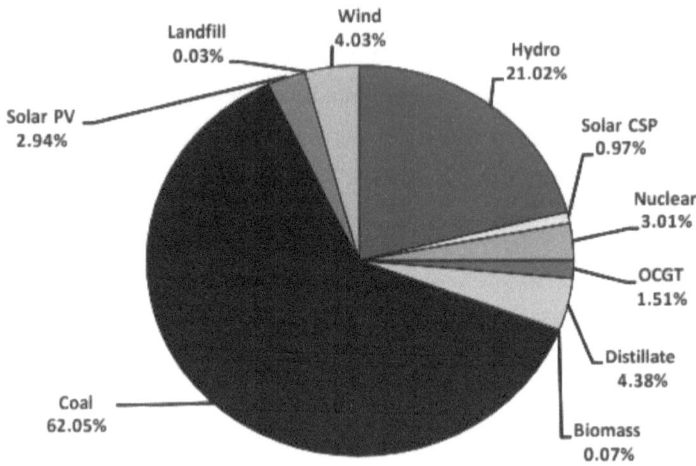

Figure 4: 1 Installed capacity by technology in SAPP countries 2017. Source: SAPP (2017).

The Intergovernmental Panel on Climate Change (IPCC) was established by the United Nations Environment Programme (UNEP) and the World Meteorological Organization (WMO) in 1988 with the aim to provide scientific information about climate change, its' impact, and potential strategies that can be adopted in response to it (IPCC, nd.: online). This led to the formation of the United Nations Framework Convention on Climate Change (UNFCCC), which was tabled in 1992 at the United Nations Conference on Environment and Development. The stated objective of the UNFCCC is to achieve stabilisation of the concentrations of greenhouse gases in the atmosphere at a level that would prevent dangerous anthropogenic interference with the climate system. The South African Government ratified the UNFCCC in August 1997. Upon the recognition that the commitments set out in the UNFCCC were inadequate for achieving its ultimate objective and after extensive international negotiation, the Kyoto Protocol was adopted in 1997. The South African Government consented to the Kyoto Protocol in July 2002, and reinforced their commitment by preparing an Initial National Communication to the UNFCCC, in accordance with Article 12 of the

Convention. Part of this included the compiling and disseminating of detailed South African Country Studies reports on a sectoral basis (UNFCCC, 2006:21).

Whilst South Africa has developed a foreign policy environment that aims to promote an integrated regional network of clean energy use and is involved in global efforts that advance clean energy use, the country's participation is not stringent. As such, the kind of FDI and economic growth that has shown to stem from clean energy use and renewable energy is at times lost to South Africa. South Africa is a non-annex I country, which means that it is not strictly required to reduce its emissions of greenhouse gases. There are no formal commitments to the UNFCCC but, where appropriate, South Africans are encouraged to make meaningful contributions within this structure. Such lax conditions remain even though the South African economy is highly dependent on fossil fuels and is a significant emitter with a high degree of emissions per capita. Indeed, South Africa ranks as one of the world's top 15 most energy intensive economies (see emission contributions by country in Figure 4:2), with a significant contribution to greenhouse emissions at a continental level (Department of Environmental Affairs and Tourism, 2005:8-10). Such a scenario acts as a kind of double-bind for South Africa – as discussed earlier, coal occupies a central place to the South African economy and to its GDP. Thus, as mentioned earlier, to interfere with that industry would be to the immediate detriment to the economy.

MtCO2

Figure 4: 2 Ranking of the main GHG emissions by country: Adapted from Global Carbon Atlas (2017).

The Department of Environmental Affairs is the main department that leads the climate change response in South Africa. Here, it is recognised that climate change is a cross- cutting issue that has a wide range of consequences that have a reach in other government departments. Therefore, effective coordination amongst the various government departments that are involved in climate change will be required to ensure that response measures are properly directed, acceptable to all, and carried out with a targeted national focus. As it stands, awareness within government on the likely impacts of climate change is somewhat limited to those departments directly involved with the issues. This has limited efforts intended to adapt to climate change and to building capacity to prepare adequately for the likely impacts. Since there are insufficient skills and capacity to formulate appropriate response measures surrounding global warming, officials do not view it as a priority and some even see it as an obstacle against the achievement of national development priorities. Instead, the concern is that South Africa has a huge backlog of service delivery deliverables where the performance of each department is measured by how effective and efficient it is on intended outcomes (Department of Environmental Affairs, n.d.: online).

Conflicting policies are also evident within the Department of Environmental Affairs as signified by the recent signing (August 2012) of a Memorandum of Understanding (MoU) between South Africa and Botswana of the Regional Environment and Social Assessment (RESA) of a coal-based energy project along the border of the two countries (Department of Environmental Affairs, n.d.: online). Climate change is viewed as an isolated matter at the expense of creating integrated departmental solutions with win-win outcomes. Therefore, climate change needs to be addressed in such a way as to assist these departments to achieve their service delivery objectives i.e. through the so-called "win-win" or "no regrets" measures (Department of Environmental Affairs and Tourism, 2005:8-10).

Since IPPs are a relatively new area of procurement in South Africa, the incumbent ANC Government will need to go through a huge learning curve in order to establish a more proficient administration machine able to deal effectively with procurement questions. According to Nel (2015:57) investors have cited an inability to achieve approvals within a reasonable period, along with deficient communication and transparency as factors that create uncertainty. This affects both foreign direct investment and the deployment of energy negatively.

4.2. Efforts towards socio-economic change

4.2.1. Trade union activism

The role of trade unions in South Africa's democratisation process is complex. The importance of trade unions in democratic South Africa stems from a history of injustices against black workers under apartheid. The role of trade unions is primarily to participate in the following broad categories: wage-setting activities, political activities, and acting as the mid-point between members and their employers (Armstrong & Steenkamp, 2008:3).

The early 1990s heralded a new phase in South African history marked by a new political dispensation that based its values on an equal society that would enjoy equal opportunities in the social, economic and political spheres. A series of legislations were passed in order to redress the past imbalances and inequalities. Some of the actions that followed ranged from the unbanning of political parties such as the ANC, Pan African Congress (PAC) and SACP to the disbanding of the South African Congress of Trade Unions (SACTU), which was to merge with COSATU, and the establishment of the Convention for a Democratic South Africa (CODESA). These actions culminated in the eventual agreement on various issues such as a new Constitution, transitional structures and a date for democratic elections (Nyathi, 2010:20).

This tumultuous period saw several initiatives aimed at correcting the labour injustices of the past, such as the implementation of strategic plans that would merge labour relations with the Constitution, which were articulated in the document from the then labour minister Tito Mboweni called "A Strategic Approach for the Department of Labour". Enshrined in the Constitution were numerous rights that would uphold fair and inclusive labour practices, including, but not limited to the right to form and join trade unions and employers' organisations, to organise and bargain collectively and to strike (Nyathi, 2010:21).

The formation of Nedlac (the National Economic Development and Labour Council) in 1995 involved representatives of the state, organised business, organised labour and communities. At the core of its mandate was ensuring a balanced and fair working environment for all and to consider all proposed labour legislation and significant changes to social and economic policy before introduction and implementation in parliament (Trade Unions, 2014: online).

The formation of the new Labour Relations Act in 1995 provided for the right to strike and organise at plant level, and created the Commission for Conciliation, Mediation and Arbitration (CCMA). The new government confronted the legacy of the apartheid regime by introducing a Skills Development Act to accelerate skill development and an Employment Equity Act to provide equal opportunities for previously disadvantaged sections of the workforce (Nyathi, 2010:28).

Of note was also the formation of the tripartite alliance (the ANC, SACP and COSATU). The fundamental objective of the tripartite alliance was founded on a common commitment to the

National Democratic Revolution: to establish a democratic and non-racial South Africa where there is continued economic transformation, continued political and economic democratisation; and the need to unite the largest possible cross-section of South Africans behind these objectives (COSATU, n.d.: online). Currently, COSATU represents the largest trade union in South Africa representing the mass of the workers in the country as members, and is able to offer a significant contribution and exert a profound influence on national economic and political policies. COSATU is vocal on its position regarding a renewable energy pathway, and the need to implement what it calls a "just transition" to such an economy. Such a transition implies putting the needs of the working and poor first, and according to COSATU, that requires an end to liberalising the energy market with ESKOM ceasing to sign any new IPP contracts. COSATU recommends that the RE sector be state owned and in instances where workers are dismissed due to the introduction of IPPs, there should be income support in addition to their unemployment insurance fund (UIF) (COSATU, 2018: online).

South Africa today is known for its activism in protest action, and according to SAPA (2013: online), a total of 99 strikes were recorded in 2012, with many of them unprotected illegal and at times violent strikes (see the amount of strikes per year in Figure 4:3). The strikes involved 241 391 workers and resulted in workers losing R6.6 billion in wages. Strike action usually occurs in essential service sectors such as electricity, gas, steam and hot water supply, health and social work, and local authority. The chief reason for strike action is remuneration and feelings of exclusion in the creation of wealth and supported with usually a weak dispute resolution framework (Le Roux & Cohen 2016:6). Moreover, trade union rivalry and the fierce competition to attract and retain members often leads to staging strike actions, which are militant and that petition for economically unfeasible demands instead of progressing employee welfare within an economically sustainable business environment (Murwirapachena & Sibanda 2014:555).

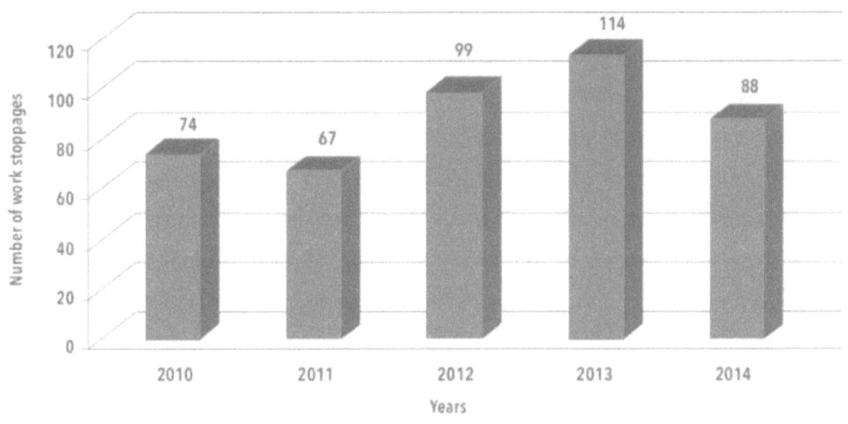

Figure 4: 3 Labour strike incidents. Source: Department of Labour (2014).

According to the Department of Labour (2015: online) the period between 2005 to 2015 has witnessed an annual average of 85 strike incidents, 5.2 million working days lost and 335 000 workers involved per annum respectively. Such strike action represents the kind of socio-political upheavals, social risk, as well threats to safety and security (in the case of violent strikes) that Venter (2005) identifies as factors that heighten a country's political risk profile and thus act as detrimental to FDI. Again, though, the situation is complex. Although unrest in terms of labour conditions as expressed through strike action represents socio-political upheaval, it also represents a drive towards change in terms of government policies, which challenges staleness in government as a feature of heightening political risk (Venter, 2005). In other words, trade union strikes satisfy one of Venter's (2005) requirements for a favourable political risk profile (efforts towards social change) and misses the mark on others (socio-political upheaval and threats to safety). As such,

South Africa in this case measures both favourably and unfavourably against Venter's (2005) political risk model.

Today, the largest modern trade unions in South Africa are: COSATU, the National Council of Trade Unions (NACTU) and the Federation of Unions of South Africa (FEDUSA). Combined, these union federations represent millions of workers across the country. While the role of trade unions in South Africa pre-1994 related to the political struggle for democracy, including the democratisation of the workplace, today trade unions can be seen as actively involved in other domains such as matters relating to economic inequality, inadequate social welfare, food and energy prices, and the eradication of e-tolls (see the principal causes comparison of two years in Table 4:1) (Gordon *et al.*, 2013: online).

Table 4: 1 Attribution to labour strike action in 2016. Source: Department of Labour (2016).

Principal cause	Quarter 1	Quarter 2	Quarter 3	Quarter 4	Total 2016	Total 2015	Change (%)
Wages, bonus and other compensation	107 034	36 819	445 041	189 980	778 874	697 810	11.62
Working conditions	13 370	40 656	1 361	5 360	60 747	26 226	131.63
Disciplinary matters	80	268	4 400	0	4 748	52 460	90.95
Grievances	4 624	17 177	19 832	9 249	50 882	43 922	15.85
Socio-economic and political conditions	650	376	1 353	6 001	8 380	9 448	11.30
Secondary action	0	23	1 362	0	1 385	2 812	50.75
Retrenchments\ redundancy	1 668	591	100	0	2 359	4 145	43.09
Refusal to bargain	5 574	2 765	0	13 012	21 351	7 228	195.39
Trade union recognition	205	6 457	0	9 800	16 462	55 624	70.40
Other factors	155 095	0	0	0	1 135	0	
Total	**134 340**	**105 132**	**473 449**	**233 402**	**946 323**	903 921	4.69

4.2.2. Labour policy

The most prominent representation of the welfare challenges facing post-apartheid South Africa is in the economy's unemployment rate (see South Africa's unemployment rate percentage by year in Figure 4:4). South Africa contends with the irony of being formally labelled as an upper-middle-income country while possessing one of the highest unemployment rates in the world (see South Africa's unemployment rate compared to OECD countries in Table 4:2). This characteristic has placed market regulation high on the agenda of pertinent policy issues in South Africa. A combination of the complex nature of these issues and the fact that the society is characterised by strong, vocal trade unions and employer associations, has meant that changes to the regulatory and institutional framework is a highly contested policy issue in South Africa (Bhorat & Cheadle, 2007:1).

There are several reasons for the increase in unemployment. Firstly, South Africa has been subject to the same skill-biased technological changes as many other parts of the world. This has hit especially hard in the mining and agriculture sectors, precisely where many unskilled blacks worked (Levinsohn, 2007:1). At the same time, the huge influx of mostly under-educated black women into the labour market has been met by a decline in demand for less skilled workers. This increase in labour supply, coupled with a decline in labour demand, has led to wage declines and a substantial increase in unemployment (Levinsohn, 2007:1). Incidentally, RE implementation attracts semi-skilled individuals, and given a surplus unskilled labour force, this feature represents a risk factor for FDI in the RE sector as per Venter's (2005) model.

This challenge has placed the South African economy with the triple challenges of poverty, unemployment, and inequality. It is noteworthy that among the three challenges, high unemployment (27.7% in quarter three of 2017) coincides with low economic growth and a recession of the South African economy during which the GDP decreased by 0,7% during the first quarter of 2017. This followed a 0,3% contraction in the fourth quarter of 2016 (StatsSA, 2014: online).

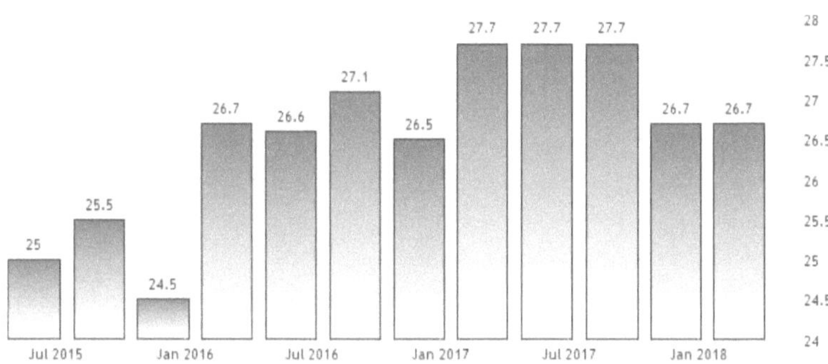

Figure 4: 4 SA unemployment rate (%). Source: StatsSA (2014).

In response to these difficulties which present issues for South Africa's political risk profile as per Venter (2005), a plethora of legislation has been formulated by the current government: the Labour Relations Act (1995); the Basic Conditions of Employment Act (1997); the Employment Equity Act (1998); and the Skills Development Empowerment Act (1998). It has been the task of the state to transform the skills regime by moving the system away from its apartheid "low skill" roots towards a framework based on free market regulation, reviving the apprenticeship system, and creating a new institutional environment structured around industry training. The kinds of high level skills mentioned here would be promoted by a functioning RE sector. Coal mining in nature makes provision for mass employment but offers jobs that are low paid and often unskilled. RE, on the other hand, would promote higher level skills jobs that contribute more directly to the economy and earn higher wages. These jobs would be fewer, however, unless there is consistent drive towards the development and advancement of a robust RE sector.

Table 4: 2 Unemployment rank in OECD countries compared to South Africa. Source: OECD 2019.

Country	Total % unemployment of labour force
Australia	5.224
Austria	4.620
Belgium	5.516
Canada	5.633
Chile	6.811
Colombia	10.327
Czech Republic	1.960
Denmark	4.866
Estonia	5.078
Finland	6.751
France	8.438
Germany	3.070
Greece	17.383
Hungary	3.411
Iceland	3.435
Ireland	5.174
Israel	3.857
Italy	9.938
Japan	2.367
Korea	3.500
Latvia	6.336
Lithuania	6.145
Luxembourg	5.314
Mexico	3.495
Netherlands	3.316
New Zealand	3.900
Norway	3.349
Poland	3.475
Portugal	6.488
Slovak Republic	5.678
Slovenia	4.341
South Africa	28.690
Spain	14.132
Sweden	6.465
Switzerland	4.338
Turkey	13.842
United Kingdom	3.765
United States	3.633

The drive toward industry training and apprenticeship was pioneered in the form of a split between the Department of Education (responsible for schools, adult education, colleges, and universities) and a Department of Labour that was in charge of "skills development" (Allais, 2011:2). Additionally, The Department of Labour (1997) introduced a national Skills Development Strategy to replace the apartheid "skills" system. Sectoral Education and Training Authorities (Setas)[2] were set up to replace the Industry Training Boards.

Additionally, the National Skills Fund was intended to fund training for disadvantaged groups, particularly the unemployed. In a similar fashion, there was a phasing out of the apprenticeship system, to be replaced by "learnerships", which would be "demand-led", in the sense that they would be offered in response to social or economic needs, including, but not limited to the formal sector's needs (Allais, 2011:3).

The South African government has produced many policies that intend to address the unemployment condition. The aim of these policies is to complement the already existing legislative and regulatory environment. For instance, there is the broad plan named AsgiSA and the Broad-based Black Economic Empowerment initiative. The shortfall in these policies is that they are unable to reach and affect the younger and the less educated segments of the population (where the majority of unemployment exists) (Brynard, 2011:73). The Expanded Public Works Programme (EPWP) is another South African government intervention that aims to address unemployment. The EPWP can potentially employ large numbers of poorly educated and otherwise unemployed workers however, it is short-term in nature and provides only temporary job creation (Brynard, 2011:74).

Criticism about South Africa's labour market mainly surrounds the vocal nature of its trade unions, which has caused the real wage to outperform productivity or output. This sort of misalignment leads to the very job losses unions seek to prevent, exacerbating South Africa's unemployment crisis and influencing firms to prefer the option of employing machinery to human capital. South Africa is reported to have higher labour costs compared to other middle-income countries such as Brazil, China, Poland, and others. Furthermore, South African salaries must contend with the ever-rising inflation rate causing further pressure on wages (Munyeka, 2014:129). These conditions increase the country's risk profile according to (Venter, 2005) and thus decrease the country's

---

[2]Setas are stakeholder bodies, with employer and trade union representatives on their boards. They were set up through a levy-grant system, through which employers pay 1% of payroll costs, 80% of which goes to the Seta. The Setas distribute grants back to employers upon receipt of training plans and reports. The hope was that this would create an incentive for employers to train and supply information that would build an understanding of the training needs of each sector (Allais, 2011:2).

global competitive attractiveness for FDI. High labour costs are unfavourable to attracting FDI, and are additionally an unfavourable feature as per Venter's (2005) political risk model.

Over and above the shortage in identifying or recruiting professional and technical staff, the South African labour laws which were in part designed to protect workers from unfair practices, inadvertently make it difficult to dismiss incompetent workers, with an array of lengthy and often difficult-to-implement rules and steps that employers have to follow before they can prove incompetence (Meibo & Peiqiang, 2013:36 – 37). This is likely to contribute to incompetence in the workplace. For the concern of energy, particularly relevant here would be Eskom and the extent to which incompetent or corrupt employees are protected in a manner detrimental to the attraction of FDI.

4.2.3. Black economic empowerment in South Africa: background

As part of bringing equality to the country, the ANC government introduced the Black Economic Empowerment (BEE) Act which is meant to economically empower the black majority previously marginalised under apartheid (van Wyk, Dahmer and Custy, 2004:265). This Act is based on the understanding that a sizeable middle class is the anchor to South Africa's democracy, and thus political stability. The South African government has furthermore embarked upon a broad-based BEE policy that has the aim of promoting economic empowerment and black shares in business ownership (van Wyk, Dahmer and Custy, 2004:265). At its core, BEE prescribes the way that business is conducted, relating to factors such as equity ownership; executive control employment; affirmative procurement and supply; the transfer of skills; corporate social investment in disadvantaged communities; and the development of entrepreneurship for small and medium enterprises (SMEs).

Gelb *et al.* (2007:6) explains BEE by describing it by its stages. In the first stage, BEE involved white companies selling a proportion of their unissued equity to a few pre-selected black purchasers. The sales were financed by loans that were often provided by the seller and usually secured by future earnings. In many instances, the buyer was a consortium consisting of one or two black individuals, usually with a high political profile but with limited experience in business. This stage of BEE deals was characterised by little government intervention until the period of 1996 when the government became actively involved in the promotion of affirmative action legislation and black enterprises in various forms.

The second stage of BEE promotion emerged during the stock market decline of 1998, which influenced the untying of financial deals. In 1999 the government was displaying greater support for BEE transactions by creating institutions such as the BEE Commission under the chairmanship of Cyril Ramaphosa. This was to be followed by a broadening in definition and constitution of BEE – including aspects of human resources, employment equity enterprise development, preferential procurement, as well as investment, ownership and control of enterprises and economic assets (Gelb *et al.*, 2007:7).

A part of the objectives of the Commission was to achieve a reformation of the South African economy through the following goals: transferal of at least 30% of productive land to black peasants and collective organisations; an increase of the black equity participation in the economy to 25%; 25% black ownership of JSE-listed shares; 40% of non-executive and executive directors in JSE-listed companies being black; 50% of government procurement being directed to black-owned companies; 30% of private sector procurement being directed to black-owned companies; 40% black executives in the private sector; 50% of the borrowers from public finance institutions being black-owned companies; 30% of contracts and concessions made by the government involving black companies, and 40% of government incentives to the private sector going to black companies (Gelb *et al.*, 2007:8).

Instead of cultivating a black middle class where small businesses thrive, BEE has produced a small class of wealthy black elites, that have come to be called "tenderpreneurs" who through lucrative public deals and the accumulation of personal fortunes, protect a false sense of economic empowerment since it promotes the capitalist economics whose foundation is the maximisation of profits (Hirsch 2005:2018). As such, BEE has been associated with nepotism and corruption. Also, these perspectives argue that the creation of a black elite class has been at the expense of the black working class population as well as black entrepreneurs and small business owners. In other words, the narrative of criticism against BEE is that it has benefitted a concentrated group of people and has given rise to corrupt practices. Since BEE has indirectly contributed to corruption, its implementation contributes to what Venter (2005) refers to as administrative incompetence and legitimacy of government as features that heighten risk profiles and discourage FDI.

BEE, however, remains an important strategy in correcting racial imbalances of the past as it also promotes social responsibility and the empowerment of communities (Juggernath *et al.*, 2011:8224). BEE is not only a moral initiative to redress the wrongs of the past, it is also a pragmatic growth strategy that aims to realise the country's full economic potential as it includes the black majority into the economy. For example, companies are given BEE status according to the extent they have included economic ownership or participation of blacks in the business.

Companies with a higher status will benefit more from public sector work and procurement policies. Conversely, a company with a low BEE status, will have less access to the mainstream activities of the government of the Republic (Juggernath *et al.*, 2011:8224).

The dominant narrative from the government, however, is that BEE acts as a conduit for foreign firms wishing to invest in the country. In 2013, then President Jacob Zuma, announced that there had been an estimated recording of R600 billion in BEE transactions since 1995. This is particularly important when the target of the ruling party is to create equity within the economy (Paton, n.d.: online). It is the belief of the ruling party that BEE will ultimately contribute to substantial economic transformation and black empowerment (van Wyk *et al.*, 2004:266).

4.2.4. Implications of the above variables for the renewable energy sector

South Africa defines itself as a developmental state (Swilling *et al.*, 2015:4-11). The main feature of developmental states is that of structural transformations and modernising economies. The country aims to achieve these ideals through industrialisation as a lever of accelerated economic growth that raises GDP per capita and creates employment. This is achieved by stimulating economic growth, and by increasing public investment in national infrastructure to stimulate private sector co-investments. The country fairs poorly in this respect and has instead created non-developmental welfarism through the issue of social grants to more people than those participating in formal employment. Consequently, South Africa possesses one of the lowest growth rates in Africa (Swilling *et al.*, 2015:4-11). For Venter (2005) welfarism and the socio-economic conditions that it produces that hamper growth, would lead to a higher risk profile and hesitance in terms of FDI.

South Africa has also used various discourse to call for sustainable energy transition notably in the Energy White Paper (1998) as well as The White Paper 2003. Additionally, Trevor Manuel (former Minister in the Presidency and Chairman of the National Planning Commission (NPC)) speaking to the National Assembly in June 2011 put forward that South Africa is the "[*27th*] *largest economy in the world, but [they] produce more carbon dioxide emissions than all but eleven countries in the world*" and that "[*South Africa is] a water scarce country but we use our water inefficiently*". He also said that South Africans "*have to change these patterns of consumption and [they] have to learn to use [...] natural resources more efficiently*". However, the most significant obstacle to addressing these threats is the preservation of the MEC: a constellation of mega firms involved in energy intensive production and extractive industries, with up-and downstream partners in the manufacturing sector. The MEC also has a firm grip on the country's energy trajectory (Swilling *et al.*, 2015:11).

The policy reforms proposed in both White Papers did not anticipate the breadth and complexity of the challenge of coordinating the various efforts of varied stakeholders including Eskom and the municipalities. The municipalities successfully resisted the 1998 White Paper's electricity distribution industry (EDI) reform, until these efforts were finally abandoned in 2010. Still in 2003, the 1998 White Paper plans to unbundle Eskom were mitigated, with the state-owned enterprise (SOE) claiming that this would undermine its financial viability. Trade unions and organised labour also supported Eskom's continued monopoly, and just as these stakeholders have expressed in media statements today, believed privatisation would prompt escalating consumer prices and job losses (Hermanus, 2017:62). As the energy security outlook of the country worsened, the DME in 2006 granted Eskom a license to build Medupi, the first new coal-fired power station in more than two decades. It was set to be the largest in the world at 4,800 MW (Hermanus, 2017:62). Crucially, Medupi (and Kusile) was established at the expense of IPP development. Such a move represents South Africa's resistance to building a robust RE sector and through this its rejection of the potential economic growth that it might bring.

The attraction to a coal-based economy is also explained in the decisions made for the country's energy pathway during the first liberation Presidency of Nelson Mandela in which the basic powers of the apartheid socio-political regime were left intact. The political settlement re-arranged power relations to create a black-middle class through, for example the development of policies such as BEE, and used debt-funded instruments and financial mechanisms that simply replaced white with black officials as part of promoting a new racial nationalism agenda (Swilling *et al.*, 2015:16). While a black middle-class was created, this was at the expense of the poor and the potential for large-scale job creation. Furthermore, the power of the MEC was left unfettered, thereby foregoing the development of industrial diversification required for a developmental state, and advancing a sustainable energy transition (Swilling *et al.*, 2015:16).

Nonetheless, South Africa does boast a renewable energy sector, which was made possible by the actions of the powerful National Treasury (NT), the Department of Energy, and by powerful business interests worried about the security of supply and South Africa's image as a high carbon emitter. A concession from the World Bank which funded a large coal-fired power station (Medupi) was made based on: the acceptance of a grant to fund RE; SA's positioning within the climate change negotiation space; the build-up of expertise of debate; research within private, civil and academic niches; and of course, the rapidly declining global price of RE technologies as compared to those of coal and nuclear energy (Swilling *et al.*, 2015:17).

4.2.4.1 Renewable energy playing field

Renewable energy is used to foster the fusion of developmental and environmental objectives. Furthermore, in an effort to minimize the effects of the triple challenge, policy shaping in the RE space has made it compulsory for developers to deliver on certain community upliftment targets. Government places a weighting on bids on a 70/30% split based on price and socio-economic development (SED) and enterprise development (ED) plans respectively[3]. Over and above top

---

[3] McDain (2016:16) writes that "According to the State of Renewable Energy in South Africa report of 2015, investment into the renewable energy programme equals R193 bn over all bid windows (Bid Window 1-4), of which 28% (R53 bn) is foreign investment (debt plus equity) (DoE 2015a). South African private financial institutions financed 45% (R57.8bn) of the total IPP investment, while state owned institutions (SOEs) and development finance institutions (DFIs) cover 22% of the total project costs.. [...]Over bid window 1-4, R19 billion has been committed to socioeconomic development with R6 billion committed for enterprise development".

developmental objectives, energy policy also targets social objectives such as job creation, localisation, and the economic empowerment of previously disadvantaged groups in South Africa. The localisation potential (and associated employment benefits) constitutes one of the criteria against which scenarios are evaluated (Montmasson-Clair & Ryan, 2014:54).

The playing field is by no means equal for renewable energy projects when compared to those of other energy sources. Whilst the REIPPPP is reliant on the private sector for delivery, a design flaw emerges in the programme where government has imposed stringent conditions that require project developers to fund community development projects (McDaid, 2016:25-26). These non-price factors count 30% of the total bid valuation, where price counts for the remaining 70%. While investors find the local economic development component high, often confusing in its structure, seen to be changing between the bidding rounds, and without much guidance on how such plans are to be prepared, South African trade unions on the contrary have been vocal to express that the economic development component is too low (Eberhard *et al.*, 2014:24). Ironically, the same SED requirements are absent in Eskom' pending Medupi and Kusile power stations. It is the view of most IPP developers that in an attempt to reduce the levels of the triple challenge, government has shifted its responsibility of service delivery and community development onto the private sector. Moreover, government has a plethora of national plans specifically designed to uplift depressed communities such as the aforementioned NDP, and at local government level, the Integrated Development Plans (IDP). IPPs are skilled at generating electricity and have performed exceptionally in this respect, but are not experienced or skilled in state functions such as the implementation of service delivery at a local community level (McDaid, 2016:25-26).

Such demand upon the IPPs has led to instances of unequal power relations that occur when large, often international, corporations interact with vulnerable poor local communities leading to unfairly negotiated deals and exploitative practices. There have been some instances (not all, such as the case of the Tsitikamma community wind farm described later) where project developers initially communicate with the selected communities, promising jobs and community benefits only to spend money on goodwill projects. Hints of these practices can be seen in the secrecy of REIPPPP in being fully transparent about implementation of their community projects. Consequently, this has strained relations between the DoE IPP office and local government (McDaid, 2016:9-14 and 25).

Unsurprisingly, a 2014 review of the REIPPPP found serious shortcomings in the manner in which socio-economic benefits were rolled out. Discrepancies emerged such as in the shortfall in the quantity of the implemented community trusts, and improper consultation and communication with community members regarding the potential benefits of the projects. This was especially true regarding job creation, given that projects are limited only to a number of construction jobs (McDaid, 2016:12).

If South Africa accelerated its RE sector through reducing the proportion of coal generation, it would lead to an immediate short-to-medium period of job losses. However, while a transition to RE and a move away from a coal-intensive economy might initially see an increase in unemployment, by 2030 the cumulative development of renewable energy will have reached a size big enough to offset those initial job losses in the traditional energy sector. Thereafter, a steady increase in job numbers in the entire energy sector could be expected (Energiewende, 2016: online). China can be regarded as a case study for such increase in job creation and employment. In that country in 2014, employment reached 1.6 million after a 4% year-on-year increase. This growth was attributed to the increased global demand for PV, the growth in domestic installed capacity, and the role of China as the world solar manufacturer. Thus, while it has not been the only factor, RE has proven to contribute to the strength of China's economic growth.

By the same token, as noted earlier, if the coal-centric nature of South Africa's energy sector should continue indefinitely, the country would miss FDI and employment opportunities associated with a thriving RE sector. It is partly the short-medium term period of jobs that concerns organised labour and trade unions. In a 2012 review, the DoE released a fact sheet on potential jobs that would be created in Window 1 of the REIPPP. Information was attained from bidders based on construction and operation jobs that would be yielded by their respective project. According to the DoE's figures drawn from successful bids, the first window would create a total 23 883 jobs – 13 069 in construction and 10 814 in operations   Disappointingly, only 463 jobs would go to South

Africans, which are considered ongoing and more valuable than construction jobs (McDaid, 2014:36).

Moreover, certain criteria have been put in place where accredited South African welders are further requested international qualifications which they do not possess and are resultantly denied jobs. Alternatively, locals are tested and given a job if they pass an international test, despite the absence of some type of preparatory training. In other instances it is reported that nepotism, favouritism, and informal processes are used where community leaders are tasked to identify potential workers (McDaid, 2014:42).

Looking ahead, it is the structure of South Africa's resources sector (organised in such a way that a shortage of green skills delays reformations in conventional mining and production practices) that amplifies the stagnation of major coal employers and blocks a green evolution and its associated cumulative labour potential, and the FDI it may attract. The resource sector, such as the coal industry is built on a traditional idea of a value chain, made up of singularly defined activities (research, exploration, production processing, transportation, rehabilitation) linked activities (governance, water, energy and waste management) and sideways value chain (such as research, and beneficiation) feeding into local and regional socio-economic plans. Green skills are needed in all these activities. They are needed both as specialist skills or occupations (such as environmental management) and as additional understanding and competence within traditional mine occupations (such as site managers). This combination of skills is scarce, as is the ability to plan in an integrated manner across systems including water, energy, mining, transport, and community development. In terms of stalling progress of a green revolution, there is also the resistance from influential mining companies of renewable energy integration to consider (Rosenberg, Visser & Cobban, 2015:7).

4.3. Overriding/existing conditions

4.3.1. Racial, ethnic, and religious cleavages in South Africa

In the South African context, racial diversity can be understood along the lines of economic growth, wherein polarised societies are prone to competitive rent-seeking behaviour. This often leads to conflict in the distribution of public goods as redistributive policies are determined along ethnic lines (Fedderke *et al.*, 2004:2). Historically, a major attempt was made to socially and politically divide the country along racial lines. Blacks, whites, coloureds and Indians were separated geographically, politically, and economically. Socially, according to Fedderke *et al.* (2004:13), South Africa is divided by race, but also by language and tribal differences.

South Africa currently has 11 official national languages and has inherited a complex system of racial classification from the apartheid era. The state of language identity in South Africa, and particularly that of white South Africans, can be explained by the outbreak of the two Boer Wars (1880 – 1881 and 1899 – 1902) between British troops and Dutch settlers over territorial conflicts (de Kadt, 2005:12). A victory for the English was sealed with the 1902 Treaty of Vereeniging, bringing most of what is now South Africa under English control (de Kadt, 2005:12).

This victory also brought about an increase in wealth and education for English whites, who most often resided in urban parts such as the Cape and Natal. By contrast, Afrikaners were more likely to be situated in agrarian areas such as the Orange Free State and Transvaal (de Kadt, 2005:12).

This was to become an important feature of social and political life well into the apartheid period (1948 – 1994). During the same period, South Africans were organised into four racial groups: blacks (about 80%), whites (about 9%), coloureds (about 9%) and Indians (about 2%). Under the separatist designs of white rule during this era, the majority black population was further sub-divided into 9 ethnic groups (Zulu, Southern and Northern Sotho, Venda, Xhosa, Tswana, Ndebele, Swati, Tsonga that would occupy their designated homelands, which were far flung from the cities and made up only 13% of the entire Republic. This was done so that the white population (including white Afrikaners) would be the most dominant and most united group in the country (Yoichi, n.d.:10).

4.3.2. Socio-economic conditions in South Africa

Despite its middle-income status, South Africa faces numerous socio-economic challenges that if not addressed could endanger the country's long-term stability. The ANC government has made efforts in tackling the legacy of apartheid to ensure that the majority black population, who were previously consigned to the margins of the mainstream economy, are brought into it. It is a daunting challenge compounded by the dearth of skills, poor education for the majority of black children, and poverty and its attendant consequences.

In an effort to redress past imbalances, the post-1994 ANC-led government has undertaken extensive policy reforms. The numerous policies created by the government have been inspired by the fundamental goals of building a united and prosperous society. The world events during this time, such as the changes in Central and Eastern Europe, the growth explosion of the Pacific Rim, the economic structural adjustment approach of the World Bank and the International Monetary Fund (IMF), the neo-conservatism of the US and Britain, and the globalisation and regionalisation of economies were inevitably also able to exert their influence in South Africa (Kotze, 2000). Moreover, these influences expressed themselves in two distinctly defined ideas of development and related philosophies of democracy: on the one hand, growth (meaning profit maximisation), and on the other hand, development and redistribution (meaning social equity) (Kotze, 2000).

In the policy shaping space, the Growth, Employment and Redistribution (GEAR)[4] was replaced in 2005 by the Accelerated and Shared Growth Initiative for South Africa (AsgiSA)[5]. AsgiSA intended to accelerate the growth of South Africa's economy, as well as accelerate wealth redistribution. However, AsgiSA was soon replaced by the New Growth Path (NGP),[6] which was introduced in 2010. The NGP focuses on job creation, the need to create decent work and a new policy orientation towards labour-intensive approaches. It aspires to grow employment by 5 million by 2020 and reduce narrow unemployment by 10%, largely through a public infrastructure

---

[4] Gear was a macroeconomic policy framework formulated in 1996 to stimulate economic growth contributing to social investment. It was also intended to reduce fiscal deficits, decrease barriers to trade and maintain the exchange rate (sahistory: online)

[5] AsgiSA's outline was to halve unemployment and poverty; improve state efficiency and minimise the regulatory burden on small–medium enterprises (SMEs) (MDG, 2013:17).

[6] The NGP's major aim was job creation (MDG, 2013:18).

programme. This was followed by the 2012 National Development Plan (NDP)[7] or Vision 2030 of the National Planning Commission (NPC). NDP is less a policy document and more a consensus-building mechanism towards a state in which poverty, inequality and unemployment will have been drastically reduced (Gumede, 2013:2).

As a middle-income, emerging-market economy, South Africa is characterised by a sophisticated financial system and the maturity of its capital markets. Sound macroeconomic and fiscal management have improved the country's position, as well as re-establishing global economic links that were sanctioned during the time of apartheid (JSE, 2009:2). Ironically, in an article discussing South Africa's "reconciliation barometer", Everett (2012:7) highlights that South African unemployment figures constitute 72 per cent of youth. This is the same category of the population that is at the forefront of service delivery protests and acts that can lead to social upheaval, risking social stability in South Africa. The biggest challenge facing South Africa is bridging the widening gap that prevails between the rich and poor, and making sure that the reforms undertaken post 1994 translate into a better life for all. GDP per capita in South Africa is far below that of the world's wealthiest nations at 75004.30 USD in 2016 (trading economics, n.d.: online). The GDP is weighed down by income inequalities as evidenced in an income per capita Gini[8] index of 0.68% in 2015 and the existence of individuals living in poverty (55.5% of the population in 2015 (StatsSA, n.d.: online) The HIV prevalence rate is approximately 12,6% among the South African population of 56,52 million, and 18% of individuals aged 15-49 are HIV positive (StatsSA, 2017: online). The poorest 10% of the population contribute 1.3 per cent of the country's income, while the richest 10% of the country contribute 44.9%of the country's income. All together these factors construct the country's human development index, which places South Africa at position 119 out of 188 countries (UNDP, 2016:2).

---

[7] The NDP's outline has been to eliminate poverty and reduce unemployment, improve school education, deconstruct the spatial patterns of the apartheid system, reduce unemployment from 27% to 14% by 2020 and to 6% by 2030, decrease the level of inequality as measured by the Gini coefficient from 0.7 in 2007 to 0.6 in 2030, become a less resource-intensive economy and adopt sustainable development practices (MDG, 2013:18).
[8] The Gini index measures the extent to which income distribution among the individuals of an economy deviates from a perfectly equal distribution. A Gini index of 0 indicates perfect equality, and an index of 100 indicates perfect inequality (World Bank, 2014: online).

4.3.3. Macro-political economic factors

Venter (2005, 46 – 49) measures South Africa's macro-political circumstances against the following variables: income tax; structural problems in the economy; some general macroeconomic indicators; and the ability to attract foreign direct investment.

In terms of these variables it is observed that South Africa possesses some of the symbols of a BRICS (Brazil, Russia, India, China, South Africa) nation, such as a large population and emerging middle class. South Africa, however, struggles to close the inequality gap (an inheritance of apartheid), and continues to face an arduous public health problem in the form of HIV Aids and tuberculosis, low average level of education, which inevitably translates to skill shortages and high levels of unemployment of 26%. South Africa's growth potential is additionally abridged owing to an inadequate infrastructure network and insufficient power supply, reflected in a low level of gross fixed capital formation (investment) of only 19% of GDP (Hammarlund, 2012:2).

Compared to its counterparts in Brazil, Russia, India, and China, it has been shown that labour costs are an important determinant of FDI (Munyeka, 2014:129). It is due to this component that South Africa has been attracting far less FDI and exporting less industrial output than in its peer group. High labour costs also contribute to a high political risk profile as per Venter (2005). South Africa is characterised by trade unions demanding salary hikes more than double the rate of inflation, and this has the impact of raising labour costs, with low manufacturing productivity (Munyeka, 2014:129).

Munyeka (2014:129) cites the example of China, also established as a global leader in RE, which has an abundant supply of human capital of 1.3 billion people and a labour participation rate of 59 per cent. South Africa, on the other hand, has 49 million people and had a labour participation rate of 36 per cent in 2010. Unit labour costs have also been higher for South Africa than in China. Additionally, South African managers earn nine times as much as unskilled workers, compared to about twice as much in China. While 80% of South African labourers reported that they had not received any training, between 60% and 80% of unskilled and skilled workers in China reported having received training. This enhances the quality of labour and skill abilities provided in China, and as a result, the country sees an inflow of multinational corporations (MNCs) wanting to build relationships with the state. South Africa's FDI net inflows as a percentage of GDP are still low (0.37%) compared to China (3.12%).

It is also interesting to note, that Chinese authorities have considered attracting FDI as an important goal, as it fosters the introduction of new technologies and helps to develop the export sector. Furthermore, the cumulative effect of FDI is influenced by the presence of other foreign investors as they perceive the presence of other investors as a positive signal (Tseng and Zebregs, 2002:10). These, and other tools are capitalised upon in China, influencing the tremendous adoption of RES, its' direct contribution to GDP growth, resulting in China installing more new renewable energy capacity than all of Europe and the rest of the Asia Pacific region (IRENA, 2014:3).

4.3.4. Implications of the above variables for the renewable energy sector

The above-mentioned overriding conditions describe a South Africa whose segregationist policies not only polarised its citizens, but through sanctions, also isolated the country from the rest of the globe. This heightened dependence on the mineral energy complex and has led to the use of coal to support its activities, as well as electrification of civilians according to politically defined priorities. Dependence on coal as a means of energy production and electrification has produced a culture in South Africa of protectionism and advancement of the mineral through carefully designed fossil fuel subsidies and indirect financial mechanisms made available to fossil fuel producers.

 The Overseas Development Institute calculated the value of direct fossil fuel subsidies to be an average of ZAR 213 million per year in 2013 and 2014 (Project 90 by 2030, 2017:10). Subsidies range from exemptions from environmental protection measures, income and price support, exemption from taxes, and benefiting from government procured goods and services. Other research from the University of Cape Town (UCT) Energy Research Centre (ERC) have calculated the value of direct transfers to have been between ZAR 5–23 billion per year (nominal), while foregone government revenues have been between ZAR 27 million and ZAR 3.7 billion per year (since 2007) (Project 90 by 2030, 2017:10)). Additionally, Eskom is exempt from accounting to air-quality standards on the grounds of the inability to afford the ZAR 200 billion it would cost to comply with the standards (Project 90 by 2030, 2017:10). This indirect subsidy to Eskom and exemption from air-quality standards has resulted in estimated health costs of over ZAR 230 billion over the non-compliance period and this non-compliance has been the indirect cause of over 2 200 deaths per year (Munyeka, 2014:129).

Renewable and clean energy by nature would have less of a damaging effect on the environment, but more importantly a shift away from coal dependence would also soften the culture of protectionism and subsidies that Eskom currently enjoys. Amongst the definitions, protectionism refers to measures that discriminate against foreign firms or other commercial interests (Kommerskollegium, 2016:8). The culture of protectionism can be traced back to South Africa's legacy of a closed economy where the principal source of monopoly came from protectionist policies (OECD, 2003:12). Protectionism with the use of subsidies has the undesired effect of raising business costs, inviting retaliation to change, it excludes industries and society from the benefits of globalisation and indeed FDI, and damages wealth and welfare in the domestic country (Erixon and Sally, 2010:14). To challenge the dominance of Eskom through the development of a renewable energy sector would be to break up the culture of protectionism and the distribution of electricity against politically defined priorities. However, as recognised by the National Treasury, while Eskom and municipalities rely on electricity sales as a revenue source, alternate sources of revenue streams must be considered to meet this goal. As an increasing number of higher income households go offgrid, municipalities' ability to subsidise Free Basic Electricity will diminish, which itself will have welfare impacts. However, this situation can be mitigated by installing solar PV panels on social housing, which would provide free electricity for indigent households as well as an additional income from selling power back into the grid (National Treasury, 2019:20). Not only is there an opportunity for municipalities to broaden their revenue stream by investing in renewable strategies, government would be additionally shifting away from coal dependence as a way to justify political decisions which heighten South Africa's risk profile and detract from FDI potential. However, South Africa remains resistant to such a shift due to the fear of job losses in the mining industry that would stem from a move away from coal and, as will be noted later, corrupt practices in government relating to the development of nuclear power.

Also, the subsidies and exemptions granted to Eskom place an unfair advantage to the REIPPPP programme where successful bids are split 70% based on price, and 30% based on SED. Little account is taken considering the local commercial benefits of a RE project, when in comparison foreign investors have access to subsidies in their countries of origin that enable them to compete powerfully on price (Martin & Winkler 2014:7).

Moreover, there is significant resistance of renewable energy efforts from the mineral energy complex, which constitutes some of the most influential corporates and is a collective major employer and contributor to GDP. Furthermore, there is apparent interest from the South African government as documented in the 2010 IRP to procure 9.6 GW of nuclear power at an estimated ZAR 1 trillion (Project 90 by 2030, 2017:10). Also, similar to the fossil fuel industry, the nuclear industry attracts state subsidies, with ZAR 599 million for the year 2016/17 going to the Nuclear Energy Corporation of South Africa. Historically, subsidies were channelled into the Atomic Energy Corporation, the Council for Nuclear Safety and the former Nuclear Development Corporation, which received 70–80% of the Department of Minerals and Energy budget during the 1980s. Ironically, the state mothballed the Pebble Bed Modular Reactor (PBMR) after spending an estimated ZAR 12 billion on the technology. Such commitment to the nuclear industry is concerning and perplexing especially amidst rapidly falling RE costs (which would attract FDI) and the rapid success of the REIPPPP. The REIPPPP has demonstrated the ability to procure, and deliver, renewable energy to the grid quickly and efficiently. The Programme has evolved into an extremely competitive process, owing partly to the continued decline in the price per kilowatt-hour of renewable technologies (WWF, 2014:13). As the Council for Scientific and Industrial Research (CSIR) points out, new solar and wind are 40% cheaper than new coal (with none of the same externalities) (Project 90, by 2030 2017:10). Moreover, the REIPPPP is classed as a champion programme by the international clean energy stakeholder community, and a model for public-private partnership for development of critical infrastructure projects (Bischof-Niemz *et al.*, 2016: 20).

 Under these circumstances, it is likely that corruption and ulterior interests are the reason why the state continues to pursue costlier, more polluting forms of energy production rather than RE. Unfortunately, high-level energy planning in SA seems to be directed more by patronage politics than logical and pragmatic solutions (Project 90 by 2030 2017:10). In this way, the state is acting in a manner that is detrimental to heightened FDI in the long-term. Particularly, the development of nuclear energy at the expense of RE and the corruption involved in that space raises concerns of the legitimacy of government and administrative competency in government as two key areas of Venter's (2005) political risk model.

Furthermore, with the current electricity supply structure, municipal sales of electricity represent a major income earner for municipalities with the 2014/15 national average at approximately 28% derived from electricity sales, second only to grants and subsidies at 31% (Project 90, by 2030 2017:9). This means that electricity sales subsidise a number of other services essential to low-income households and, as income from electricity sales to municipality's decrease, the resources to provide subsidised services and funds for public infrastructure maintenance will erode. This is an aspect of welfarism, and a risk factor for investors. The situation for municipalities, however, will be aggravated as the cost of rooftop solar photovoltaic (PV) technology drops and wealthy households and commercial consumers abandon a centralised electricity supply system, and start investing in their own generation facilities. Such large scale privatisation of RE sources would mean that the state would not directly benefit from the sector. When welfarism is so pronounced in the country, a transition from the current coal-intensive energy model will be challenging to disrupt (Project 90, by 2030:2017:9).

Macroeconomic considerations are important to consider in the electricity supply sector. Besides its core energy mandate (to contrast the cost of generation pathways, to consider security of supply) the IRP must also consider its overall impact on macroeconomic conditions in South Africa. This includes bearings on the exchange rate, foreign currency, and trade flows especially. The initial IRP 2010 report depicted a constrained long-term growth path, with a matching energy demand requirement. This feature was improved in the 2013 update through the designing of an electricity plan based on an aspirational 5.4% per annum growth rate. The update also prioritises a high degree of flexibility, incorporating more renewables so as to favour decisions of least regret (Montmasson-Clair & Ryan, 2014:52). Nevertheless, challenges in the energy sector (which include the inherited imbalance of access from the apartheid system, as well as the subsidies and exemptions granted to Eskom) persist.

Development of the IRP is influenced by the actions and decisions of the Electricity Supply Industry (ESI)[9] and IRP models based on the price effect of the different energy sources. This process is skewed in the sense that the ESI has secured relationships with select heavy electricity users offering them favourable long-term rates (Montmasson-Clair & Ryan, 2014:5).

The main beneficiaries of subsidised electricity are large mining and industrial consumers (such as paper and petrochemicals, and metal smelting companies) who consume 60% of South Africa's power. This very subsidisation promoted the inefficient use of energy consequently leading to the country ranking amongst the highest global GHG emitters (Deloitte, 2017:49). Furthermore, this results in harnessing a sustained dominance of the heavy users, acting as a barrier to competition to new entrants and indeed renewable energy partakers. The attainment of these unreflective rates have stirred questions of fairness and have attracted attention from the Competition Commission and Competition Tribunal of South Africa, where these actions have been labeled a violation of dominant positions by incumbents (Montmasson-Clair & Ryan, 2014:5). Subsidisation also takes place when Free Basic Electricity is given to indigent households, when, as mentioned in Section 4.4.4 this situation can be mitigated by installing solar PV panels on social housing, which would provide free electricity to these households as well as an additional income from selling power back into the grid (National Treasury, 2019:20)

Such actions reinforce the monopoly status of Eskom, making it difficult for IPPs to compete. In other words, since Eskom is the single buyer, initiatives like the REIPPPP are put at risk. Also since the risk to the REIPPPP hinges on Eskom's financial soundness, a destabilisation of Eskom's status and government's international credit rating would also impact negatively on government funding available to IPPs and indeed the REIPPPP itself (Martin & Winkler, 2014:7). This means that the development of a robust and continuous RE sector in South Africa is highly complex. Such development would challenge the Eskom monopoly and lead to heightened employment and lower costs (in the long-term) despite a challenge to Eskom also implying a decline in its funding structure. Given this scenario, South Africa faces intrinsically difficult conditions for the

---

[9] The ESI is made up of Eskom, municipalities and Independent Power Producers. Institutional stakeholders of the ESI include National Treasury (NT), Department of Energy (Doe), Department of Public Enterprise (DPE), and the National Energy Regulator of South Africa (NERSA).

development of its RE sector. In other words, the kind of catch-22 inherent in Eskom's monopoly and the importance of coal mining to the South African economy render the establishment of a robust RE sector difficult. The challenges, in other words are structural. As such, the kinds of benefits associated with RE in the country (lower costs, challenge to monopoly, FDI interest) are prohibited not only by the actions and decisions of government, but also by the make-up of the energy sector itself.

Infrastructure readiness and investment

In South Africa, Eskom has acknowledged that it is currently experiencing challenges to connect renewable energy projects to its already congested and ageing national grid (Hassim, 2017 13-14). The current Eskom grid is much more difficult to access and more often requires infrastructure investment in the form of a new substation and or new lines for new connections. This problem is a common feature in countries that are integrating renewable energy technologies into the system and occurs because current infrastructure is old and the coal combustion technology used in the electricity transmission grids is unsuitable (Hassim, 2017 13-14).

Alignment between generation and transmission planning and implementation remains an issue that requires a variety of forward planning initiatives, especially since integrating renewables is challenging, with exception of biomass, due to its intermittency (IRENA, 2016: online). What is required to overcome such deficiencies and promote the smooth integration of electricity from renewable energy is the implementation of smart grids. More so, the scalability and roll out of smart grids will gain from the infrastructure multiplier effect, which can be considerably beneficial to the economy. In certain scenarios, the multiplier effect can result in economic growth of ZAR 2 for every ZAR 1 spent on infrastructure. If the government policies incentivise infrastructure spending, the effect on growth in the economy could be boosted, resulting in much higher growth rates than those currently experienced in South Africa (Hassim, 2017 13-14). However, the combination of the ageing infrastructure of the current grid, the dominance of coal in the energy sector, and the drive toward nuclear energy acts as a hindrance to such growth rates and the heightened FDI that they would bring.

Instead, to overcome this challenge, Eskom encourages renewable energy projects to apply for a "self-build" option. The self-build option is elected in the case where it is more cost efficient and faster to self-build substations and connection lines using Eskom's specifications. This is often preferred by IPPs over risking delays in reaching commercial operation in Eskom's connection of the project to the national electricity grid (Hassim, 2017 13-14).

Unsurprisingly, South Africa lags behind its emerging market peer counterparts in terms of investment expenditure as a share of gross domestic product. Data for 2012 for example, shows that in China and India investment levels were between 1.8 to 2.8 times that of South Africa. Yet interestingly, the rates of return on investment in South Africa are not low: real returns have averaged around 15 per cent between 1994 and 2008 while nominal returns were 22 per cent between 2005 and 2008 (Bhorat *et al.*, 2017:11). Notably these rates of return are the same as that of China, albeit over a longer period. Typically, investment levels are high if returns on investment are high, but this has not been the case in South Africa. What this suggests is that non-price factors have affected the level of investment in South Africa. These range from factors of production market distortions to structural concerns around political stability and governance. In other words, the rate of return in South Africa is artificially inflated as it guides the investor and information seekers at large, influencing investment levels in the economy.

Exacerbating the problem, is South Africa's notoriously low savings levels (fueling a consumption-driven economic growth trajectory) which implies that the country relies on short-term capital flows to finance a narrow band of capital-intensive sectors in an environment of low savings, and high internal rates of return. The dependence on short- to medium-term capital inflows tends to perpetuate dependence on the resource sector and powerful oligopolies in the services sector

that provide generous margins that portfolio investors seek. As a result, just as the REIPPPP is assessed strictly on socio-economic credentials such as job creation, once low hanging fruits are taken up in the form of available substations and transmission lines, it is incumbent on IPPs to invest in supporting infrastructure in ensuring the realisation of projects. IPPs will need to achieve this while government policy shaping is conducted on a job-starved, capital intensive growth trajectory, where poverty, when measured using the official national poverty line (as updated in 2011), increased from 31 per cent in 1995 to 53.8 per cent in 2011 (Bhorat *et al.*, 2017:11).

4.4. Interrogation of the efficacy of government

4.4.1. Legitimacy of government in South Africa

The emergence of the ANC government after the 1994 national democratic elections was followed by the party almost immediately gaining the upper hand, as the demographics of the country (i.e. the black vote) appeared to guarantee the ANC a majority victory. Since then, Mattes (2007:3) states that the goal of the ruling party has been specifically that of transforming South African society, both politically and economically, as well as reshaping social life as it was known. With such a mammoth task at hand, it would be important for the ANC government to instil a great sense of patriotism and gain the commitment of the entire nation. This was attempted, in part, through the concept born in this era of the "new South Africa". The "new South Africa" strives for a unified commitment to democratic rules, an extensive trust and respect for the institutions embedded within the regime and an active participatory citizenship. Furthermore, there was recognition of the need to acknowledge a kind of strength in diversity, and allocate equal status to the vast cultural, religious and linguistic identities that are shared by South Africans. In this way, the ANC government was carving for itself a legitimate rule. Indeed, the nature of South Africa's political transformation made it clear that political institutions could not survive simply through coercion (Mattes, 2007:4).

It is interesting to note that even during the apartheid regime there existed democratic and legitimate institutions such as Parliament, political parties, political rights for whites, and an independent judiciary – all structures that legitimised the ruling NP. These structures, however, were implemented in a way that promulgated racial inequalities, favouring the white population (Broch-Graver, 2005:6). This begs the question: where do the sources of a legitimate government lie? To answer this question, Broch-Graver (2005:32) describes three types of authority that give rise to legitimate governance: legal rule (where legitimacy is based on the law and administration of the state), traditional rule (where legitimacy is based on inherited power) and charismatic rule (where legitimacy is based on the leader's personal character). It is the charismatic source of legitimacy that Broch-Graver (2005:33) describes as the most relevant to the ANC ruling government, as it usually involves the leader needing to prove his or her authority through the achievement of some form of success, such as for example, the attainment of liberation. If the performance of the leader declines, then the legitimacy of that leader will decline. This point is especially relevant in South Africa, where the divide between the rich minority and the poor majority is severe, and largely a function of apartheid and of South Africa's history of systemised racial segregation. The ANC government's legitimate authority can be seen to be eroding, which has been depicted through several protests and strikes in the country, owing to a lack of service delivery and frustrations stemming from high unemployment levels.

In former President Mbeki's second term in public office (in office between 1999-2008), allegations of the corruption of his deputy Jacob Zuma surfaced, leading to Zuma's dismissal and an internal revolt for him to be re-instated. This event was to backfire for Mbeki when he was overturned and voted out of the ANC leadership before the expiry of his full term in 2009, with accusations of his leadership style lacking consultation and often being centralised. Jacob Zuma came to represent the image of a leader with a concern for the people, and one to bridge the gap with the masses, a characteristic often perceived to be lacking in Mbeki (Suttner, 2014). To regain the hearts of the masses, Jacob Zuma's campaign was given the slogan "Working together we can do more" (Suttner, 2014). It is during the same period in 2014, during preparation for his election manifesto, that a damning report by Public Protector Thuli Madonsela revealed President Zuma's awareness of the inflated spending of taxpayers' money (over R200 million) in upgrading his private home in Nkandla. Questions arose about the legitimacy of the ANC, and it became clear that, along with the passing of Nelson Mandela, the party could not enjoy the same prestige and legitimacy enjoyed by its veteran leaders, as their credibility was based on trust and their deeds (Suttner, 2014). The concerns here about the efficacy of the ANC government and the legitimacy of government emerge as political risk concerns according to the Venter's (2005) model.

4.4.2. Safety and security in South Africa

South Africa must contend with a high crime rate not unconnected to the legacy of discrimination and economic marginalisation under apartheid, under which the majority was confined to the fringes of the mainstream economy. It is for this reason that South Africa's state of crime is often described tautologically as violent crime considering the overwhelming violence and callousness that accompany it. The country's legal system is relatively independent of political meddling, and so acts as a bulwark against lawlessness, be it in the political realm or in society generally. This is significant because strong institutions such as the judiciary are indispensable to democratic consolidation, as they ensure safety and security for all.

This indicator or variable is also sometimes described as "law and order". Law entails an assessment of the strength and impartiality of the legal system, whereas order relates to an assessment of the popular observance of the law of a country (Fouché, 2003:36). Venter (2005) uses this indicator to reflect on crime rates in South Africa, as well as the measure of corruption in society. Given the high crime rates and extent of corruption in terms of accusations against the Zuma administration, South Africa's risk profile in this regard is high. Logically, such a high risk profile creates difficulties for FDI.

South Africa's transition from an apartheid state to a constitutional democracy led to the creation of the Constitutional Court. The Constitutional Court symbolised an institution of change, entrusted with the mandate of advancing the reform of the South African jurisprudence, thereby ensuring that the law, including all legislation and common law, are aligned to the Constitution (Department of Justice & Constitutional Development, 2012: online).

In South Africa, The Bill of Rights and legislation flowing from it paved the way to establishing a legal environment that is adversative to corruption. The culture of secrecy and authoritarianism that existed before 1994 has been replaced by legal norms favouring democracy, accountability and transparency. Legislation such as the Promotion of Access to Information Act and the Promotion of Administrative Justice Act, together with institutions such as the Competition Commission and the Chapter 9 bodies make it possible for civil society and concerned citizens to hold the state and private bodies to account in ways that would have been inconceivable in the past (CASAC, 2011:8). Protected by the right of freedom of expression under Section 16 of the Constitution, the media now play an important role in the detection and exposure of corruption. Additionally, institutions such as the Asset Forfeiture Unit and the Special Investigations Unit have had several notable successes in apprehending and prosecuting corrupt individuals and recovering the proceeds of crime (CASAC, 2011:8). The South African Constitution, which is the bedrock that lays the foundation for an open society recognises the injustices of the past; upholds democratic values and promotes social justice based on an inclusive diverse nation, where every citizen is protected by law, through a freely elected government (Department of Justice, 1996:1). South Africa further belongs to four major conventions of the UN, AU, OECD, and SADC, compelling the state to actively combat corrupt activities, prevent organised crime, promote whistle-blowing, and enforce punitive measures (Corruptionwatch, 2013: online).

The World Justice Project, which commits itself to the rule of law globally, has recently released its latest Rule of Law Index, which includes 97 countries, of which South Africa is one. According to their measurement, there is noticeable improvement being made by South Africa in entrenching the rule of law in the governance of the country. These scores place South Africa in the mid-range of global scores and toward the upper end of regional rankings (Hoffman, 2013: online).

Such a shift in the Constitution and South Africa's legal and policy commitment to equality and the erosion of, encourages FDI, especially in the South African context where the new Constitution, characterised by the promotion of equality, ran parallel to the official abolishment of apartheid and the re-entry of South Africa into the global economy. Thus, politically, South Africa has a history of making decisions and executing actions that position the country favourably in terms of FDI. However, questions regarding the ANC under Jacob Zuma in terms of corruption and state capture, as well as upholding the monopoly of Eskom have gone towards weakening the political risk profile and endangering FDI. There have also been failures in the advocacy of the rule of law (see Table 4:3 & Figure 4:5). The Council for the Advancement of the South African Constitution (CASAC, 2011:1) defines corruption as the "antithesis to democracy and the rule of law" and further points out that corruption is not found only in government structures, but also in the private sector. There is a recent emergence of evidence of large-scale fraud and anti-competitive collusion by big businesses in many sectors of the South African economy. While this behaviour is a negative feature for the political risk profile in terms of Venter's concept of legitimacy of government and administrative competence, the emergence of this information is also an indication of the robustness of South Africa's regulatory bodies such as the Competition Commission[10].

Table 4: 3 Top 10 countries for economic crime. Source: PWC (2018).

Country	Economic crime (%)
South Africa	**69**
France	68
Kenya	61
Zambia	61
Spain	55
United Kingdom	55
Australia	52
Russian Federation	48
Belgium	45
Netherlands	45

[10] The Competition Commission is a statutory body formed under the Competition Act, No 89 of 1998, and amended by the Competition Second Amendment Act, no 39 of 2000 by the Government of South Africa. Its role is to investigate, evaluate, and control repressive business practices, abuse of dominant positions and collusion, thereby advancing fairness and efficiency in the South African economy (Competition Commission South Africa, 2019:online).

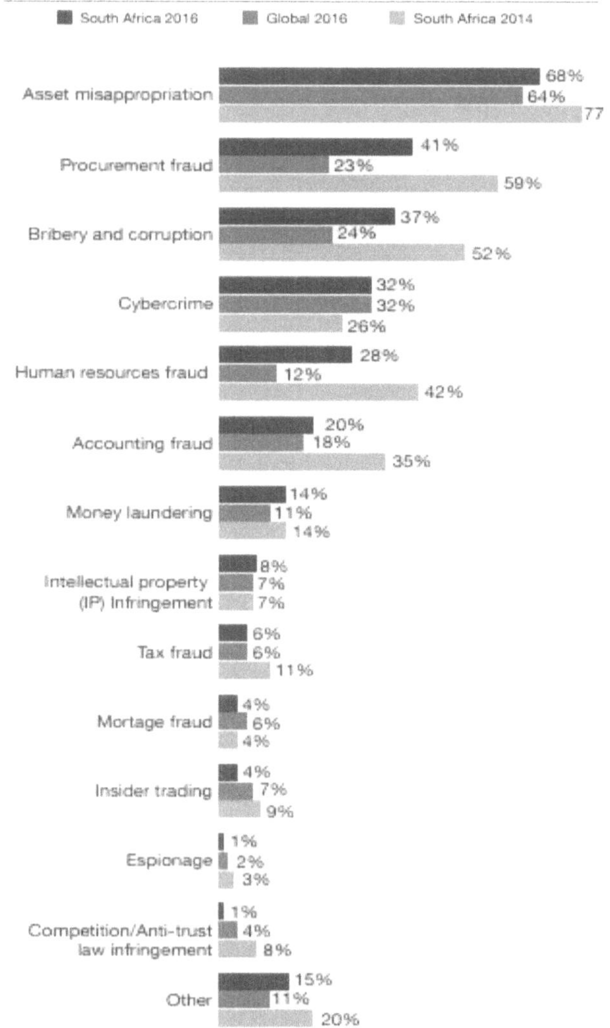

Figure 4: 5 Types of economic crimes over 2 years. Source: PWC (2018).

Business has been observed appointing politically connected candidates to senior positions or as their BEE partners in order to secure government contracts (Gumede, 2017:online). Leading banks, such as ABSA and Investec were accused in 2016 of unethical behaviour by manipulating financial markets (van Schalkwyk, 2017:27). Evidence of corruption is perhaps most notably witnessed in the government space, with a plethora of cases of government officials acting in breach of the law, often dismissed under secretive circumstances. Much of the recurrent corruption takes place at the sub-national level in government in the form of abuse of resources at municipal ranks (van Schalkwyk, 2017:27). Indeed, a number of high-profile cases relating to fraud and corruption against politically connected officials have either vanished, or have been withdrawn by the National Prosecuting Authority (NPA) (Tamukamoyo & Newham, 2012: online). At the helm of power, President Jacob Zuma has been accused of 783 charges of corruption, money laundering and racketeering (Tamukamoyo & Newham, 2012: online). Such allegations interfere with the level of FDI that the South African Constitution would otherwise promote and work towards heightening risk in South Africa's political risk profile as per Venter (2005).

4.4.3. Administrative competence in government in South Africa

The service delivery protests that South Africa has experienced in the townships (apartheid designated black settlement areas) over the lack of amenities such as electricity; water; schools; tarred roads; and even jobs partially expose the dysfunction in the local tier of government, which can be attributed to patronage. This patronage sees the appointment of incompetent but highly connected individuals (in the party-speak, "deployed") to senior positions (Venter, 2005). The consequence is the mismanagement of resources through, a genuine lack of skills (inability to spend monies), the hiring of personnel on the basis of nepotism, and the entrenchment of careerism and factionalism, which render these local units as dysfunctional sources of sinecures and personal enrichment. Judicious management of state resources at the local level, the hiring of competent individuals, and a general professionalisation of the state to decouple it from the ANC are prerequisites to addressing the myriad of social economic and political challenges that result in service delivery protests, in which lives are disrupted, public property such as schools destroyed and even lives lost. These, logically, are safety concerns that contribute to South Africa's level of social risk as a significant factor in political risk (Venter, 2005). Their prevalence in the South African space, therefore, would go towards threatening FDI.

In the Republic of South Africa, the government comprises national, provincial and local pillars, which are distinct, but interdependent and interrelated. Two leading documents have been at the centre of restructuring local government in South Africa. These are the Constitution and the White Paper on Local Government Transformation, 1998. In terms of Section 40(1) and (2) of the Constitution, local government is one of the spheres of government that seeks to promote and enhance cooperative government (Sibanda, 2012:1).

At the heart of the White Paper on transforming public service delivery was the concept that came to be known as 'Batho Pele', a Setswana phrase meaning 'people first'. The purpose of coining this phrase was to articulate the strategy of putting public services in place that were effective, people-centric and sustainable (Raga & Taylor, nd.: online).

The elevated status of local government and the associated participatory rights of citizens have led to many outbreaks of protests against poor service delivery by municipalities. Despite the tremendous strides made in democratising local government, poor service delivery and perceived corruption still result in some municipalities being dysfunctional. To be effective in monitoring service delivery, local governments both nationally and internationally are adopting New Public Management (NPM), private-sector or market-oriented practices (Sibanda, 2012:2).

The impact of fraud has led to the declaration of special regulations and improvements in existing legislation that have led to the creation, among others, of the Directorate of Special Operations, commonly known as the "Scorpions" (abolished in 2008); the Asset Forfeiture Unit; the Public Protector; the Special Investigation Unit; Commercial Crime Units; Internal Audit Units; Special Investigation Units within departments; and the appointment of forensic consultants (Ambe & Badenhorst-Weiss, 2012:251).

Managa (2012:3) summarises the administration challenges of the present government as being owed to severe problems of institutional capacity. As mentioned in Section 4.5.3, this refers to a lack of expertise that leads to many municipalities being ineffectively staffed and has resulted in poor service delivery over the years that has left many communities with inadequate access to basic services. This skill shortage exists mostly in managerial and technical positions, which remain vacant in most rural municipalities, creating overwhelming backlogs that prevent efficiency and are felt by poor communities (Managa, 2012:3). These kinds of institutional issues and the service delivery problems that they present have historically provoked strike action, thus increasing South Africa's political risk profile in two ways: first, administrative incompetence as a key area for political risk is heightened, and secondly, socio-political unrest and upheaval as a similar key area to political risk (Venter, 2005) is provoked.

While some municipalities lack adequate funds to carry out their constitutional mandate to improve service delivery, some instead underspend their allocated funds because of a lack of critical skills required to operate municipalities. These skills are project management, financial management and general leadership skills. The result is incomplete projects or failure to start others (Managa, 2012:3 – 4).

The government's mandate to deploy ANC comrades to public office positions limits access to qualified staff and worsens the prevalence of unsuitable incumbents that are unable to perform their duties successfully. Further to that, some of these political appointees enrich themselves, earning astronomical bonuses on the backs of collapsing municipalities (Managa, 2012:3 – 4). In such cases, the legitimacy of government under Venter's (2005) risk model is brought under question and FDI as a result is hindered.

There is also a ripple effect, where the lack of quality service provision leads to unsustainable budgetary management. Communities refuse to pay for inferior service, and cite a lack of affordability due to unemployment as a reason for non-payment (Managa, 2012:4).

4.4.4. Staleness of incumbency and leadership succession

According to the Institute for Security Studies (ISS) (2007), the general shift towards democracy has brought with it the entrenchment of term limits (usually ten years) that the ANC respects. Staleness of incumbency relates to the hegemonic position of a party, and in South Africa this applies to the ruling ANC, which has been in power since 1994 (Neethling, 2014:40).

The transition to democracy has seen the ANC consistently receiving majority of the electoral support (currently just under 50 percent), and it has ultimately become the dominant party in South Africa. Its symbolic association with the anti-apartheid liberation movement has largely carried it through elections with staggering margins. Wieczorek (2012:28) speaks about the concept of democratic consolidation, which is generally viewed as the ultimate end goal of democratisation, where democracy becomes "the only game in town". Here, established institutions and democratic practices become deeply ingrained within the society to the point that operation outside of them is inconceivable. The destiny of a country is not intertwined with that of the individual in power, as is the case during one-party and military dictatorships, because the rule of law and institutions guide the exercise of power[11].

In a dominant party system, multiple parties compete for power, but the ruling party wins consecutive elections, largely exploiting incumbency and state resources. In South Africa, the absence of a single formidable opposition to the ruling party reduces the threat of losing power, which affects the accountability of the government and explains undemocratic acts such as looting the economy, intimidating minorities, and holding elections in which the ANC worries about only the margin of its victory (Wieczorek, 2012:28).

The rise to presidency of Jacob Zuma was expected to herald the end of the 12 years of neoliberal policies seen under Thabo Mbeki. Mbeki's macroeconomic policies (many of which were developed in the Mandela era) were thought to have led to the rising levels of township and community protests over service delivery demands. According to Dwyer (2010:10), service delivery protests reached levels that had never been experienced in any country in the world before. These strikes rendered Mbeki unpopular, and provided an opportunity for a faction of the ANC to recall him. Following the recall of Mbeki, the ANC Polokwane Congress, through the influence of the South African Communist Party (SACP) and the trade union COSATU, elected a Zuma-led faction in December 2007 to head the party. Despite his brushes with the law,[12] Zuma was viewed as "a man

---

[11] Jacob Zuma was removed from his position as President before the end of his term due to mounting scandals and allegations of corruption.
[12] In 2002, Zuma was implicated in a publicised corruption scandal in connection with his close ally Schabir Shaik, where it was alleged that Zuma used his position in the government for kickbacks through Shaik and his companies in a government arms acquisition deal.

of the people" and a friend of the workers who was best placed to address their concerns. Zuma defeated Mbeki by more than double the votes that Mbeki secured (Dwyer, 2010:10). The recall of Mbeki was followed by the resignation of several cabinet ministers allied to him.

As South Africa's head of state since 2009, Zuma has been severely criticised in the media by analysts and commentators for his poor leadership. Zuma has blamed emerging middle-class blacks, as well as whites, for consistently finding fault with his government – a situation that could partly be attributed to his involvement in a string of court cases for corruption and disreputable sexual behaviour, as well as his reputation as a leader who relies on his homespun intelligence and canny ability to mediate between people and the many factions that make up the ANC (Neethling, 2014:41).

Jacob Zuma's position as ANC leader was reaffirmed at the ANC's Mangaung (Bloemfontein) Conference in December 2012. For many commentators, the election of ANC veteran and business magnate Cyril Ramaphosa as deputy president of the governing ANC brought hope that he would inject new energy into the party and help Zuma to steer the country along a new trajectory (Neethling, 2014:41). Even in his second term of presidency, Jacob Zuma's leadership inspired intensifying criticism from most political parties.

4.4.5. Implications of the above variables for the renewable energy sector

The ANC under Jacob Zuma had several implications in the deployment of officials in government whose decisions shape the energy sector. Essentially, the non-consenting hire and fire of officials has had much to do with unaccounted for aspirations of advancing nuclear energy despite opposition to it. Adopting nuclear energy is not only too costly for the country but has the potential to derail REIPPPPP efforts and momentum. Thus, the continued interest in it reveals a measure of corruption that acts as a hindrance for FDI and as a contributor to South Africa as a politically high-

---

In 2005, Zuma was accused of raping a woman whose father was a member of the apartheid resistance movement Mkhonto we Sizwe and who served in prison on Robben Island with Zuma. Zuma was acquitted of the rape charge, and after his acquittal was re-instated as ANC deputy president.

Zuma is embroiled in a major controversy related to costly upgrades to his private home in Nkandla, KwaZulu-Natal. State security personnel recommended improvements of around R27.9m, but by 2013 it was found that costs had risen to just under R250m (SA History, 2014: online).

risk country as per Venter's (2005) model. In these cases, in which the corruption in question limits the development of RE, South Africa misses the opportunities for job creation, lowering of production costs, and a favourable impression for foreign investors associated with a strong RE sector. What this shows is that political risk factors (such as social risk, administrative incompetency in government, and (il) legitimacy of government) are not divorced from the energy sector – rather, such political risk factors in and of themselves contribute to a high political risk profile, but their bearing on the energy sector and in particular renewable energy, exacerbates that profile and further detracts from FDI opportunities.

A timeline of events

According to Gosam (2016: online) the unfolding of events in the energy sector can be summarised as follows

- Jacob Zuma was inaugurated as South African president in May 2009. In November 2009, the Guptas, a wealthy Indian family closely linked to the president, formed a new company, called Oakbay Resources and Energy Limited. The Guptas have also obtained state contracts (mainly in coal mining), leading to allegations of state capture and dubious dealings that further heighten South Africa's high-risk profile and detract from FDI.

- One month later, in December 2009, President Zuma declared at the United Nations Climate Change Conference in Copenhagen that South Africa was going to reduce its carbon emissions by 34% by 2020. His announcement took both local and international commentators by surprise, but it also revealed Jacob Zuma's nuclear ambitions and his domestic efforts towards obtaining the necessary domestic sign off.

- In March 2011 President Zuma's cabinet approved the Integrated Resources Plan (IRP2010) 2010-2030, which was a 20-year road map that outlined the mix of the country's future electricity generation, including the need for 9 600 MW of nuclear power. In the same year, Japan encountered the Fukushima Daiichi nuclear plant disaster, which raised an alert to South Africa surrounding nuclear safety.

- Subsequently, the government's IRP2013 findings were confirmed by international consultants at KPMG and Deloitte (who were commissioned by the energy department) and evaluated by independent academic experts at UCT's Energy Research Centre, the University of Stellenbosch as well as research experts at the CSIR. These bodies concluded that nuclear technology is unnecessary as viable alternatives are available. Based on findings from the CSIR, the (levelised) cost of electricity from nuclear power is 25% more expensive than new coal or solar photovoltaic, and 67% more expensive than wind. Furthermore, an approval and implementation of the 9 600 MW nuclear programme would constitute as an illegal move under the Public Sector Finance Management Act. It would also be in direct conflict with the ANC's National Development Plan (NDP) and ANC 2015 national general council resolutions.

- In a strategic move, in July 2013, President Zuma took control of the inter-ministerial National Nuclear Energy Executive Coordinating Committee (NNEECC), by replacing its Chair, then deputy president Kgalema Motlanthe, with himself. The NNEECC was soon after replaced by the energy security subcommittee.

- On 25 November 2013 the then minister of energy Ben Martins acted as witness to the signing of an agreement between Rosatom and the South African Nuclear Energy Corporation (Necsa), which gave the energy department the go-ahead to procure 9 600 MW of nuclear power from Russia. This move gave president Zuma confidence to announce an eminent sign off procuring 9 600 MW of nuclear energy in a State of the Nation address in February 2014. In May of the same year, Jacob Zuma reshuffles cabinet, and demotes Gordhan, an opponent of the nuclear deal, as finance minister replacing him with the low profile but respected Nhlanhla Nene. Zuma also appointed a new energy minister, the "political lightweight" Tina Joemat-Pettersson.

- In September 2014, Jacob Zuma appointed himself chair of the cabinet's energy security. The Subcommittee was led by the minister of energy Joemat-Pettersson. This self-appointment enabled Zuma to by-pass Eskom as the owner and operator of the nuclear fleet, excluding any technical and financial oversight by them, and also to neutralise minister of public enterprises Lynne Brown, who was sceptical of the nuclear programme.

- On 22 September 2014 a joint media release by Rosatom and the South African government announced that the two countries had signed an intergovernmental framework agreement which laid the foundation for the large-scale nuclear power plants procurement and development programme. This would be achieved by the construction of 8 new nuclear power plant units with a total installed capacity of 9 600 MW. Joemat-Pettersson endorses this move.

- In July 2015 conflict arose between the National Treasury (NT) and DoE, as NT resisted the implementation of the procurement process. NT could not justify such expenditure, deeming it unaffordable.

- In mid-July 2015 — shortly after Zuma, Nene and Joemat-Pettersson returned from the BRICS summit in Russia – the energy department announced that the procurement would start that same month. This message was reinforced in the August 2015 mid-year State of the Nation speech, where the president stated that the procurement should be concluded within that financial year. That would imply that the first reactor would be built in 2019. Stockpiling uranium before any reactors come online, would place well positioned Oakbay ahead of the queue due to Shiva Uranium's (a Gupta owned uranium mine) close proximity to Jacob Zuma.

- Meanwhile, On 27 August 2015 Nene reaffirmed that he would not sign off the $100-billion deal to build nuclear power stations, struck between Jacob Zuma and Russian President, if it was "unaffordable".

The replacement of Nene in December 2015 upon his expression of his stance against nuclear power created the impression that Zuma aimed to organise his cabinet exclusively with those who would affirm his decisions. This kind of scenario casts doubts on South Africa's level of governmental legitimacy and the strength of its administrative competence as two integral components of how political risk profiles are evaluated under Venter's (2005) model for political risk analysis.

On 14 October 2017, a letter from Minister Kubayi instructed the DoE IPP Office to proceed to conclude PPAs with the successful REIPPP Bid Window 3.5. Almost immediately following this letter, Minister Kubayi was removed as energy minister and replaced by former minister of state security, David Mahlobo without any resolve from Eskom, whether it will meet its' legal obligation to conclude Round 3.5 and 4 of IPPs above R0,77/kWh (Cape Business News, 2017: online).

Eskom has also been unwavering about its pro-nuclear stance and acts as one of the minister's few vocal supporters of nuclear expansion. In December 2016, it was announced that Eskom would be the official procurement agent for nuclear power, which it would own, operate, and finance through further significant international borrowing (Hermanus, 2017:72). This is despite its already dire financial situation. This means that the original determination, and the subject of the High Court challenge, had changed (Hermanus, 2017:72).

It is arguable, therefore, that alternative energy futures are at the heart of the South African political crisis. According to the CSIR, in 2016 the price of renewable energy was 62c/kWh over the life cycle, compared to coal which was R1.03 - R1.20/kWh and nuclear was R1.30/kWh over the life cycle (Bhorat *et al.*, 2017:17). The CSIR estimated that the nuclear energy option could result in an increased annual cost of R90 billion compared to the cost of renewable energy (Bhorat *et al.*, 2017:17). There is, therefore, no economic rationale for building nuclear power plants in South Africa. Experts argue that the most likely reason that Eskom and the Zuma-centered power elite pursue the nuclear option is because it creates an opportunity to extract rents on a massive scale while giving the Russians the strategic advantage they aim to achieve in building all their nuclear power plants. The Russians are currently building nuclear power plants in 30 countries (Bhorat *et al.*, 2017:17).

Given the country's apartheid legacy, transforming the core administrations is a crucial undertaking requiring extraordinary levels of dedication, ethics, technical capacity and a well-defined governance programme. Although significant progress has been made, there is now dissatisfaction across society and within the ANC itself with the performance of these institutions. Instead of serving the public good, with the mantra of 'Batho Pele', the state has created a shadow state onto the existing constitutional state, redirecting energies to enrichment of a small power elite (Bhorat *et al.*, 2017:4).

4.5. Concerns relating to stability for the private sector

4.5.1. Military involvement in politics

South Africa was characterised as a highly militarised, war-orientated state during the apartheid years. It was the role of the military to act as the backbone of the apartheid system, giving shape to the security of the regime nationally and further afield, in order to protect South Africa's borders (Pringle, 2010). With the transition into democracy in 1994, it would happen that the new (ANC) regime would inherit a sophisticated military machine, as well as the reputation of being the regional tyrants. Through demilitarising its image, the newly elected ANC government sought to replace its former reputation and become a nation that is committed to the conciliatory functions of promoting peace and security throughout the continent (Pringle, 2010). This was seen through the various phases of leadership starting with Nelson Mandela and his implementation of budgetary downscales, together with a closer focus on transforming the military in terms of its structure and introducing an ethos of national reconciliation.

The decision to demilitarise naturally decreases political risk in terms of Venter's (2005) appraisal of military involvement in politics as part of authoritarian governments that present high risks for foreign investment. In this regard, South Africa under the ANC has positioned itself favourable in terms of attracting FDI.

4.5.2. Social risks: extremism, religious tensions and terrorism

Ogbonnaya (2013:60) defines extremism as the sort of philosophy or act – whether religious, political, socio-cultural or economic – that is perceived to infringe on the common moral and acceptable standards that exist mostly in the political arena. In contemporary political discourse, however, the term extremism is almost invariably linked to religious activities. Religious extremism usually finds comfort in its brutal acts, as religion is used to provide justification for the violence (Michael, 2007:40). Terrorism – which refers to the intended use of violence in order to terrify certain groups of society – is an extension of extremism. It deploys calculated politically motivated violence, usually against civilian targets, through underground agents.

The issues of marginalisation, injustice, and structural violence owing to socio-economic inequalities, poverty, political exclusion, and lawlessness have the capacity to push people, especially youths, into extremist ideologies (Harris, 2002:169). Although South Africa has not suffered from extremist violence in the sense of terrorism, service delivery protests, xenophobic violence and violent crime are variants of extremism and are indicative of the socio-economic fissures in South Africa's society that need to be addressed to guarantee the country's stability, enhance democracy, and stand as an attractive country to foreign investors.

4.5.3. Effect of xenophobia on private sector

While post-apartheid South Africa carried the appearance of a country triumphant over racially motivated violence, realities such as xenophobic attacks on black foreign nationals bely what Harris (2002:169) refers to as "a new pathology for a "new South Africa'". This type of violence has typically been motivated by competition for political and economic power and involved in many instances the mobilisation of small business owners in inciting violence and evicting foreign nationals as a way to regain territory (Polzer, 2010:2).

4.5.4. The security of private property

There are twin problems in South Africa when discussing private property: that of nationalisation, and that of land reform.

Since the discovery of minerals in the 19th century, South Africa has managed to place itself firmly on the resources map, benefiting considerably from private funding invested in the extraction, refinement and export of the precious commodities. Capitalism further helped the country to modernise, improving South Africa's industrial state and bringing with it the development of railway and other transport infrastructure, the development of small towns and the emergence of Johannesburg (Forrest *et al.*, 2010: 373). Kimberley was the first town in the Southern hemisphere that had electrical street lights installed even before London could use this electrical technology (Eskom, nd.: online). It is most notably the discovery of minerals that has propelled the entire South African economy, which is estimated to be currently worth R349 USD billion (Trading Economics 2018: online).

There have been public debates about nationalising the mining sector and other strategic sectors. Such discussions can cause apprehensions among the business community at large concerning the speculative consequences that nationalisation is thought to bring (Forrest *et al.*, 2010:374).

The ANC government has not formally endorsed the prospect of nationalising any key sectors, with individuals such as Secretary-General Gwede Mantashe, and other key figures speaking against it outright. Some factions have, however, publicly shown their support for this type of policy change. The government's biggest ally, COSATU, expressed its enthusiastic advocacy in recent times of such an ideal. COSATU's former senior economist and advisor to Finance Minister Malusi Gigaba, Chris Malekane, has not only been a vigorous supporter of nationalisation, but has been quoted as saying that the trade union federation would reject any report critical of nationalisation. COSATU is only open to a discussion relating to choosing between different models of nationalisation (du Plessis, 2011:3).

There are many arguments that oppose the route of nationalisation, one of these being the efficiency argument. Quite simply, this argument claims that if the strategic sector has been efficient and competitive, then change is not necessary (du Plessis, 2011:9).

It is believed that nationalised firms also tend to have confused goals and nationalising the resource and strategic sectors will raise government debt and cost the government more than it receives, which will eventually undermine the scope of the distributive policies on the national budget. Besides discouraging potential and current investors in the country, the nationalisation debate has been labelled a solution that will merely serve the short-run political goals of its proponents at the expense of long-run economic efficiency (du Plessis, 2011:10).

South Africa's history of racial discrimination has left the country with a skewed pattern of land ownership that excludes most of the black population from land assets (DA, 2013). The legacy of dispossession and forced removal under colonialism and apartheid (Lahiff, 2001) now calls for the post-apartheid government to contend with issues surrounding land, where the state must mobilise resources to reverse the effects of displacement by previous land policies.

The current picture in South Africa is that of 13 million poverty-stricken black people remain crowded in the homelands, where rights to land are unclear. There is an ongoing exodus of people from the rural areas into the cities. In the cities themselves, shack settlements, plagued with crime and a lack of basic services, are on the increase (Lahiff, 2001). This is an example of how South Africa's imbalanced (and unjust) political history has resulted in very real challenges in the present that ultimately increase its political risk as per Venter (2005) and lower its standing as an attractive space for FDI. In these instances, the government are contending with years of social imbalance that have resulted in political risk factors such as safety, social risk, and racial and ethnic cleavages.

4.5.5. Implications of the above variables for the renewable energy sector

Despite the emergence of small-scale embedded generator projects, private sector participation in RE is mostly through REIPPP projects which require agreements with landowners as most of the REIPPPP projects are being built on commercial farmland used primarily for grazing. IPPs thus need to secure agreements with landowners on transfers of ownership or lease rights for REIPPPP projects. It has been discovered that current legal agreements and prevailing land ownership patterns privilege the interests of commercial (mainly white) landowners who receive the majority of renewable energy projects, amounting to an average 2% of the total revenues over the 20-year life of the project (McEwan, 2017:5). Again, structural features of South Africa in the 21st century as influenced by the apartheid era have significant bearing on the energy sector and the renewable energy concern in particular. In this case, historical racial cleavages have led to a centralisation of access to not only energy but to contracts involving renewable energy, to a racial minority of land-owners. Here, South Africa's history of racial segregation is shown to have bearing on not only land ownership but also on the potential of the RE sector to function on a national level. This, in addition to decisions from government regarding nuclear power as well the monopoly of Eskom and the continuing dependence of coal, acts as a significant impediment to the development of a strong RE sector.

Nevertheless, efforts have been made to address this structural challenge to the implementation of RE, some of which have been successful. Notably, renewable energy transition has coincided with new legislation regarding land rights. In June 2014, the Restitution of Land Rights Amendment

Act reopened the land claims process that closed in 1998, giving groups and individuals able to prove dispossession after 1913 a further five years to lodge claims for restitution or compensation. By March 2016, 143,720 new claims had been lodged. Though the process has been described as one presenting clashes between agriculture and mining on the best practices for land use, the Northern Cape boasts the most claims settled for land restitution to rural communities and families. This implies that IPPs will be required to include potential local land claims in project risk profiles (McEwan, 2017:5).

Under the Act, equity is enhanced for the dispossessed who benefit from lease payments accruing from the REIPPPP, or from claims for land restitution. One successful case concerns the Tsitsikamma Community Wind Farm near Wittekleibosch in the Eastern Cape, built on land where amaFengu people were forcibly removed under apartheid in the 1970s. The community returned to the land after a claim was lodged in 1994 and is now a 9% shareholder of the wind farm (McEwan, 2017:5). This kind of development balances the corruption and secretive hidden agenda-driven practices of government, with decisions that not only benefit disadvantaged communities but also heighten the scope for RE implementation and the lowering of costs and indirect FDI it would generate.

However, legislation does not help those dispossessed prior to 1913 and, and in those cases, REIPPPP could heighten perceptions of injustice and thus damage FDI as per the Venter (2005) model. For example, four neighbouring communities in uMkuze in KwaZulu-Natal recently brought a claim against Charl Senekal, South Africa's largest sugar farmer and KwaZulu-Natal's largest private landowner, whose bid to build a R1.1 billion, 16.5 MW biomass plant approved in Round 3. Whilst the claim has delayed development, it may not be successful because Senekal has counterclaimed that the land has been occupied by whites since 1880 (McEwan, 2017:5).

The important message that emerges for IPPs is that land occupation in South Africa is connected to very particular forms of territorialisation and struggle. Land rights legislation and claims involving REIPPPP projects must be handled delicately and prudently in appreciation of the historical land injustices that can weigh on the placement of renewable energy projects and greater energy transition. From government's perspective, the threat of land grab from global investors against the pursuit of attracting foreign investment with the limitation of foreign ownership of land is an imperative. In other words, there exists a tension between foreign land ownership as part of FDI and the drive to address historical imbalances in land ownership through promoting local land ownership for those who had been previously disposed in the country itself. In this particular case, FDI may act against the interest of addressing South Africa's racially segregated political history. Similarly to the challenge in addressing historical imbalance and the undesirable but almost necessary Eskom monopoly, this tension between encouraging FDI and promoting local interests represents another structural challenge to South Africa's large-scale implementation of a fully functioning RE sector.

As has been the norm, efforts have been made by the government to resolve or at least address these kinds of tensions, showing that government can act in a competent and integral manner that is attractive to FDI and beneficial to its political risk profile. In this case, in order to avert fears and reduce complications to investors, the South African government has created spatial technology zones. In terms of spatial politics these zones are spaces of some commercial activity, possessing overlapping forms of sovereignty, where international, inter-governmental and non-governmental actors come together under an administrative authority that knits together a unique legislative process (McEwan, 2017:5).

While the development of these spatial zones is intended to create a frictionless realm of legal and economic exemptions, there is undoubtedly potential for commotion in respect to competing political, social and spatial interests. Competing in the REIPPPP is permissible only once a bid has complied with the necessary environmental standards, land rights have been issued, and supporting documents such as a notarial lease registration and proof of land use applications are in place. This process can be lengthy in that several permissions are required, from the Department of Environmental Affairs which itself is inundated with a high volume of renewable energy projects (McEwan, 2017:6), a clear indication of interest by investors seeking approvals. Thus, while efforts are at times made to promote RE, the execution of the policies and plans is often arduous and can stall the process of striving toward a robust RE sector.

Furthermore, areas of friction in the geographic setting of renewable energy projects can arise between corporate and community stakeholders if the REIPPPP is perceived to be driven solely by global capital rather than by resolving the legacy of energy apartheid. Energy apartheid separates those who can afford from those who cannot, and those with access to the grid from those who do not. This can already be witnessed on the wind farms of the Eastern Cape coast, where communities adjacent to renewable energy projects are aggrieved by the lack of access to electricity despite their proximity to the grid (McEwan, 2017:9). Therefore, as desirable as RE is for environmental and economic concerns, the manner in which the development of it manifests has been shown to foster inequality through the pursuit of global capitalism. This observation supports the argument presented in Section 4.3.4.1 where it is demonstrated that there is a complex situation in which capitalism as a vehicle for FDI clashes with the concern of addressing structural inequality. Because attracting FDI is so bound up in capitalism, and because capital is centred so heavily in the white racial minority, the attraction of FDI in the form of RE is often at odds with development disadvantaged local communities.

Understandably, to respond to the needs of those 12 million South Africans who do not have access to grid electricity is beyond the scope of IPPs and as a result zones of friction are emerging from the expectations created in remote rural areas, in which the South African government has failed to deliver large scale grid electricity and access. Here, friction that often results in social unrest and a threat to safety as part of Venter's (2005) model for political risk stands as a risk to FDI. However, in cases like these, the friction and social risk is a product of historical unequal land distribution and racial cleavages which itself presents a threat to FDI.

5. Conclusions and recommendations

In Chapter 4 various socio-economic, and political dimensions were used to profile South Africa, examining the contribution of those dimensions to the country's energy sector. In this chapter, findings of the book will be used to draw the main conclusions as well as make some necessary recommendations that will aid and boost South Africa's renewable energy sector.

5.1. Conclusions

Over the past decade, a strong business case has been established for the exploitation of renewable energy sources due to ongoing falling costs and the recognition by many countries that renewables present the pathway to global energy transformation, energy security, and abatement of climate change. Furthermore, renewables can bridge the gap for energy-poor societies, thus creating opportunities for sustainable livelihoods for the millions of people who lack energy access throughout the global economy. The energy policy shaping space is influenced by the demands of the sustainability goal, associated with the complexity of the socio-economic and biophysical systems. Rising fossil fuel prices, societal impacts of energy use and observed climate change make the exploitation of renewable energy more important than ever.

Despite these desirable outcomes, there are risks involved in the application of renewable energy that have understated the continued robustness at which they can be implemented. The South African government, as it currently operates, displays obstacles to the implementation of renewable energy. A strong renewable energy sector has been shown to attract foreign investment (as has been the case in China), hence South Africa's resistance to a continued robust RE sector represents a loss of opportunity to promote and attract direct FDI and thus foster economic growth.

FDI in the energy sector, and specifically renewable energy stands as a desired outcome not only as a mechanism for climate change mitigation, but also for bridging the energy gap, and the generation of economic growth. National development policies such as the National Development

Plan (NDP) and the Draft White Paper on Foreign Policy highlight the importance of FDI for meeting South Africa's development priorities in addressing the triple challenges of poverty, unemployment and inequality.

Illustrating the significance of RE and its related FDI contribution to economic growth is the fact that between the periods of 2010-2015, R192.6 billion was invested in renewable-energy projects in South Africa. Of this, 28% (R53.2 billion) came from foreign investment and finance. This figure almost matches the entire foreign direct investment in the SA economy in 2014 (R62 billion).

By using the Venter (2005) model as a framework for political risk analysis and in respect of RES exploitation, three primary obstacles are shown to hinder the continued progress of a thriving RE sector: firstly, the coal-dependent nature of the South African energy sector and the central position of coal to the South African economy means that any shifts away from this model would yield unemployment and challenges to the economy. These, though, would be in the short-term and the risk of a move toward renewable energy would yield long-term benefits such as heightened employment and increased FDI. Secondly, the drive for nuclear energy in conversation with Russia is primarily driven by corruption and agendas not in the interest of the economy at large, has led to the stalling of meaningful and long-term renewable energy implementation. While various policies and papers on renewable energy have been drafted, and while some areas of South Africa (the Northern Cape particularly) have seen advances in the sector, successful renewable energy implementation falls victim to corruption and ulterior motivation. Thirdly, South Africa's history of racial segregation and the historical imbalanced provision of electricity poses natural challenges for equal distribution of land and access to it. Racial discrimination has left the country with a skewed pattern of land ownership that excludes the majority of the black population from land assets. The legacy of dispossession and forced removal under colonialism and apartheid now calls for the post-apartheid government to contend with issues surrounding land, where the state must mobilise resources to reverse the effects of displacement by previous land policies. In other words, access to energy in its most traditional form is already challenging for the government, and so pioneering an entirely different system may be even more so. Related to this is the structural issue of the clash between engaging in the global economy on the level of FDI and investing in developing the local community, especially in respect to disadvantaged communities who lack capital and access to electricity.

All of the above manifests in South Africa's energy sector, which is characterised by its own complexities that interfere with FDI and heighten social risk. At times, this is a product of the aversion to risk and the desire to maintain monopoly (Eskom and coal, coal as historically and naturally integral to South Africa's economy), or can be a function of corrupt practices (nuclear power and Russia), and at other times it is a result of the difficulties given rise by apartheid (historical unequal land distribution has led to current imbalance in capital and access to the grid), and finally is partly a product of the natural tension between global competition and FDI and addressing imbalances locally.

Based on the above mentioned pointers, South Africa is observed to experience friction as government navigates international and local concerns that are at odds in the advancement of a robust renewables energy sector and its implied foreign direct investment amidst an environment of limited availability of access to capital in the country.